PENGUIN BOOKS

THE FULLNESS OF WINGS

Gary Dorsey has worked as a staff writer at *The Hartford Courant's Northeast* magazine and at several other newspapers, including the *Detroit Free Press*. An award-winning reporter, he lives in Coventry, Connecticut. He is at work on a new book.

AFTERWORD BY

JOHN S. LANGFORD, PROJECT DIRECTOR,

THE DAEDALUS PROJECT

PENGUIN BOOKS

THE FULLNESS OF WINGS

The Making of a New Daedalus

GARY DORSEY

PENGUIN BOOKS
Published by the Penguin Group
Viking Penguin, a division of Penguin Books USA Inc.,
375 Hudson Street, New York, New York 10014, U.S.A.
Penguin Books Ltd, 27 Wrights Lane, London W8 5TZ, England
Penguin Books Australia Ltd, Ringwood, Victoria, Australia
Penguin Books Canada Ltd, 2801 John Street,
Markham, Ontario, Canada L3R 1B4
Penguin Books (N.Z.) Ltd, 182–190 Wairau Road,
Auckland 10, New Zealand

Penguin Books Ltd, Registered Offices:
Harmondsworth, Middlesex, England

First published in the United States of America by
Viking Penguin, a division of Penguin Books USA Inc., 1990
Published in Penguin Books 1991

1 3 5 7 9 10 8 6 4 2

Photo inserts follow pages 160 and 224.

THE LIBRARY OF CONGRESS HAS CATALOGUED THE HARDCOVER AS FOLLOWS:
Dorsey, Gary.
The fullness of wings : the making of a new Daedalus /
Gary Dorsey.
p. cm.
ISBN 0-670-82444-5 (hc.)
ISBN 0 14 01.1485 8 (pbk.)
1. Human powered aircraft. 2. Daedalus (Greek mythology)
3. Aeronautics—Aegean Sea Region—Flights. I. Title.
TL769.D67 1990
629.13—dc20 89-40666

Printed in the United States of America
Set in Bodoni Book
Designed by Fritz Metsch
Map by Virginia Norey

FOR JAN WINBURN,

WHO COULD HAVE BEEN AN ASTRONAUT

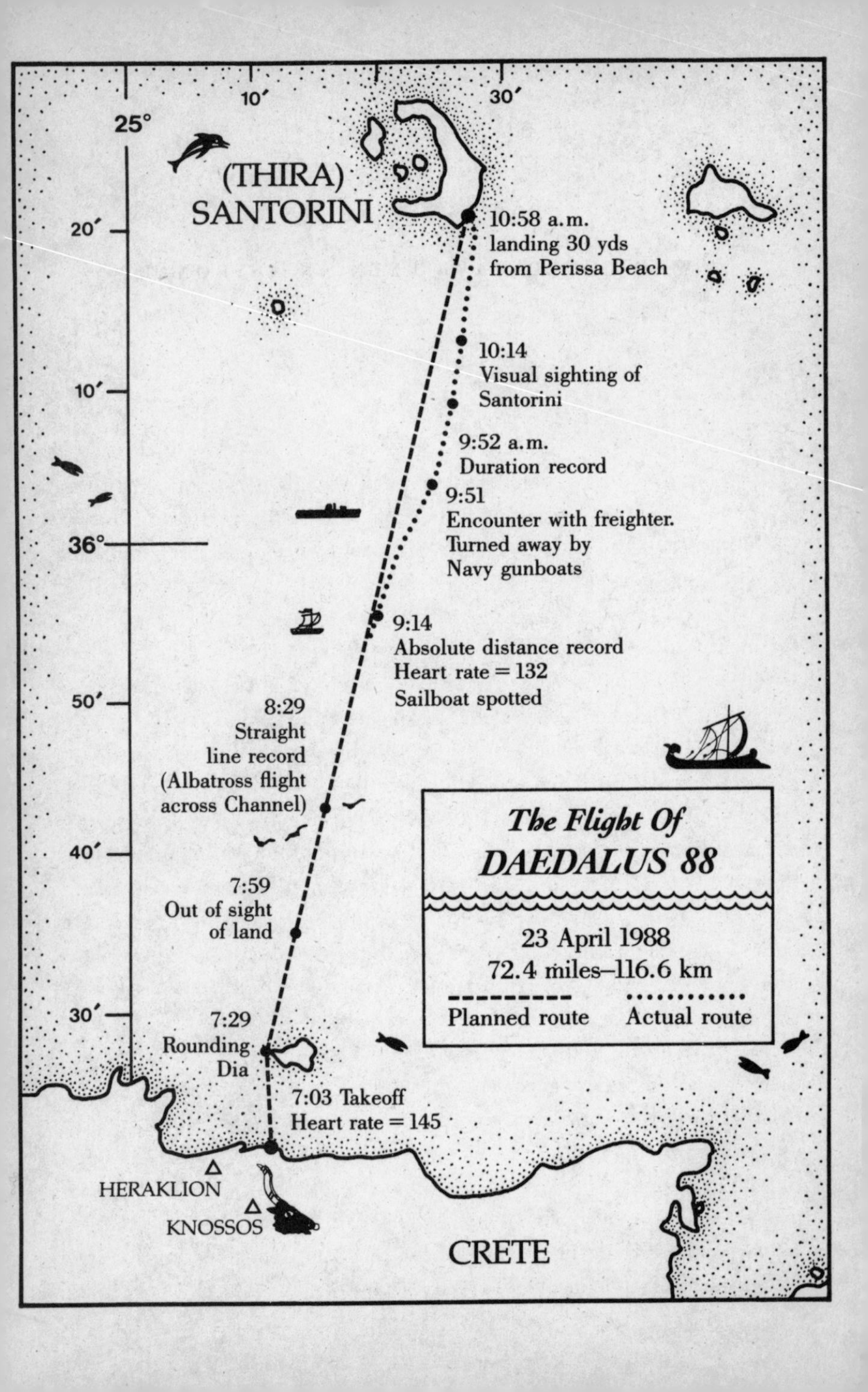

10′
30′
25°
(THIRA)
SANTORINI
20′
10:58 a.m.
landing 30 yds
from Perissa Beach
10:14
Visual sighting of
Santorini
10′
9:52 a.m.
Duration record
9:51
Encounter with freighter.
Turned away by
Navy gunboats
36°
9:14
Absolute distance record
Heart rate = 132
Sailboat spotted
50′
8:29
Straight
line record
(Albatross flight
across Channel)
40′
7:59
Out of sight
of land
30′
7:29
Rounding
Dia
7:03 Takeoff
Heart rate = 145
HERAKLION
KNOSSOS
CRETE
The Flight Of
DAEDALUS 88
23 April 1988
72.4 miles–116.6 km
Planned route
Actual route

ACKNOWLEDGMENTS

This book, like the Daedalus airplane, had a single person piloting it, but a whole team of people helping it fly. I am grateful, in particular, to Juan Cruz, Mark Drela, Bob Parks, and Hal Youngren for offering accelerated courses in aerodynamics and engineering over six months. Many of their lessons have been woefully simplified on these pages, and I can only hope they will forgive the deficiency of my interpretations.

The physiological mysteries of human-powered flight are the province of Dr. Ethan Nadel, and I am grateful for his aid to my understanding through analogies, journal articles, references, and dinnertable discussions. Glenn Tremml, besides encouraging me to enter the life of the Daedalus Project, also served as a guide through the maze of physiology. Louis Toth cleared the path many times so I could grasp the basics of pilot testing procedures.

While I relied, for the better part of a year, on my own observations, interviews, and research within the Daedalus team to write most of this book, I also depended on the diaries, daily journals, letters, and recollections of team members and associates to develop those portions of the story where I was not actually an observer. In other instances where I have recreated dialogue and scenes, I have tried to portray events only when there was either independent corroboration among the principals or when existing records of an event provided corroboration. In this respect, I am especially grateful to documentary film makers Bob Nesson and Mark Davis for supplying me with hundreds of hours of unedited videotape recordings of the early meetings of the Daedalus team, the first construction efforts, interviews with team members and flights of the Michelob Light Eagle. Like a horde of students at MIT, I am also grateful for Al Shaw's generosity.

Al's small archive of videotapes, documenting the construction and trial flights of the BURD, Chrysalis, and Monarch, were indispensable aids in my research.

I cannot find any satisfactory way to thank John Langford. I'm sure there were many times when he wondered whether he would regret having allowed a journalist to roam so freely into his life and through the hangar, invading meetings, plumbing the moods and minds of talented, sometimes critical, crew members. John not only tolerated months of intrusion, but encouraged it. The enthusiasm he showed for this book, even during the team's most painful episodes, continues to strike me as remarkable.

There are numerous former colleagues at *The Hartford Courant* to whom I feel indebted, beginning with Lary Bloom, the editor of *Northeast Magazine* who saw an opportunity for the growth of one of his staff writers and promoted the idea within *The Hartford Courant*. Thanks also to the executive editor of *The Hartford Courant*, Michael Waller, who took a gamble against tradition and gave me a year's leave of absence to bring this book to life. Of course, I never would have taken the first step without the encouragement of my peers at the *The Hartford Courant*, especially staff writer Joel Lang, who saw the characteristics of a good adventure story long before I did.

I have many people to thank for their personal support, especially those on the Daedalus team itself, whose numbers and kindnesses exceed my ability to recount here. Pilots, physiologists, engineers, UROPs, clerical workers formed a family for a while. Most importantly, my own blood ties aided this enterprise. For their interest and unwavering support, I thank my parents, Hugh and Kate Dorsey; my sisters, Barbara King and Joan McPherson; my brother John Dorsey. Their presence often buoyed this occasionally lonely business. I also owe a debt of gratitude to non-fiction writer Mark Kramer—his assistance once or twice helped me see a way through the project when the woods seemed too thick. I feel fortunate, too, because my agent, Flip Brophy, and editor, Pam Dorman, had vision enough not to fret about whether the Daedalus airplane ever crossed the Aegean. Their patience and enthusiasm throughout the course of the Daedalus adventure directed me through some stormy channels.

At last, I thank Jan Winburn. It's through her spirit that I initially came to see the value of this project and to believe that in the pursuit of dreams we experience the fullness of wings. Her energy, enthusiasm and talent have been evident to me at each critical stage, and I've happily received her many gifts. She's inspired enough lift, balance, and control to make me feel like I flew, too. I can only hope this reads like her kind of book.

PROLOGUE

In Boston they still talked about him as if he'd vanished into a cloud, leaped like a bird from their berth in Cambridge, lifted up across the Charles River, negotiated through Back Bay. Thermaled over the harbor. Out to the Atlantic. On to the Aegean. The engineers barely endured his flights back and forth from Boston to Washington, from Florida to California, from New York to Athens. They tried to ignore their globe-trotting manager, flying from nowhere to nowhere. In the past two months he'd traveled almost every other day—including two trips to Greece—and on each return he'd brought back a new emergency plan, another scheduling change, a different demand.

John Langford might as well have been Harry Houdini drumming up curiosity seekers in the last days of the nineteenth century instead of the manager of the Daedalus Project leading a new generation of radical engineers toward the end of the twentieth. His team had learned the calculations, equations, and formulas for human-powered flight, and despite what they thought, he still valued engineering over media hype. He hadn't forgotten their classical antecedent—*Et ignotas animum dimittit in artes*. (And he turns his mind to unknown arts.) But who else would satisfy the weenies? While his engineers spoke sanctimoniously of the spirit, he searched for commercial support, boats, vans, and money.

Regardless of pride and suspicions and cynicism, young hands kept sanding and smoothing 112-foot Styrofoam wings stretched across sawhorses in a brightly illuminated hangar in rural Massachusetts. Just days before Daedalus' flight date, all-night building crews wandered the hangar. They slept on concrete floors, clambered around oily milling machines, invented new tools, fed on M&Ms and sodas and pizzas. Renegades from the aerospace industry instructed students from Cambridge how best to meld sophisticated aerodynamic designs with the lightest, strongest plastics and carbon composites. Freed from uninspiring jobs and lackluster classrooms, they quietly stalked the myth of a new Daedalus with the tools a child might use to build toy models: razors, epoxy, sandpaper, balsa. They stripped down to their shorts and plotted midnight surgery to make the wings lighter and stronger. They labored under their own artifice, beneath long and lustrous pink wings as delicate as down. Late at night they entered a peculiar realm of science and imagination, the perfect and timeless enchantment.

Within days more than three dozen engineers, physiologists, athletes, and office workers would have to be in Crete. On a runway near Knossos they'd judge the texture of seawater and the caress of winds, and wait for a perfect day to make their escape. After years immersed in an engineering project to build the world's most elegant airplane—an intoxicating confluence of art and science—they would have to fly.

More than a million dollars in corporate bargains had brought them to this juncture. The public relations staff spoke wearily of their flight as "the Stunt," and Langford continued to arrange deals and search for money. A series of delays narrowed the flight window, commercial backers threatened to dissolve funding, weather forecasters predicted windy conditions that promised to keep Daedalus at bay. In the end, the winds of Aeolus, not engineers, would set the agenda.

This, of course, was the way of their myth. For thirty-five hundred years the Daedalus story had been told as a cautionary tale warning children to obey their elders, not to stray too far, not to fly too high. The flight of Western civilization's archetypal engineer had become so confused with his son's failure—the fall of Icarus—that even in Greece schoolchildren interpreted the myth as an instruction to heed authority. But they believed the world was wrong. Daedalus, a mortal who made his way working in the courtyards and temples of gods and kings, flew out of a labyrinth in Crete with feather-light wings, and though he suffered a son's death along the way, he escaped safely by traveling a realm that was not naturally human. This had been the dream and inspiration, from Leonardo to the Wright Brothers, from Kitty Hawk to Cambridge.

In March 1988, when the team arrived in Knossos, the mission sought to defy all traditional accounts of the Daedalus tale. They had challenged the gods and kings of the modern world, corporate and governmental and scholastic bureaucracies that sought to make money off imaginative aeronautical explorers. Their pronouncements attracted attention all over the world. Elements of an old, enervated story—the bliss of Icarus, the creative power of Daedalus, the winding labyrinth—would come to life again. Despite the tensions that divided them, despite the spirit of a myth whose best known version was a warning, climaxed by failure, they would attempt to fly from Crete. They pledged to revise the myth. They would write it anew.

FIRST TOYS

1

Once in the woods near Peachtree Creek, not far from the Bobby Jones Golf Course and the Atlanta Memorial Arts Center, way up on Northside Drive, a little boy's war games diminished and dwindled and, finally, disappeared. The regular blitz of toy tanks, plastic soldiers, and backyard battles played behind mud fortifications gave way to more organized games after school: the Viking Club, Cub Scouts. But he lived too far away from the others; he had allergies; he was too small. When other boys' war fantasies turned to hard scrapping football games in neighborhoods far away, the boy in the woods turned to airplanes. He went indoors.

They worried about John. Up in that room of his all alone, bowed over a desk, snapping rudders and wings and ailerons off plastic stems, squinting at a puzzling array of airplane parts, hoarding his twenty-five-cent a week allowance like a drug addict to buy more parts and tubes of intoxicating glues to build more airplanes to hang from the ceiling. He worked untold hours in an unventilated room. His father, the Judge, didn't like it. They had all that space outdoors. The family doctor didn't like it either.

"Margaret," the doctor said one day, during a house call to treat John's bronchitis—he walked up the stairs, glanced into the bedroom, and pulled her aside—"Margaret, just what are you doing to this child?"

"Well, Dr. McKee, you know how many precautions we take. . . ." Mrs. Langford stammered. She'd done everything to protect her boy's health. She baked bread for him out of rice flour rather than wheat. She made him lamb patties so he wouldn't have to eat beef. When he went to birthday parties, she sent him off with his own little slice of rice cake.

"Margaret, you know what I'm talking about," the doctor said. "How on earth are we going to treat allergies as bad as his when he has all those airplanes hanging from the ceiling?"

Mrs. Langford peered into the room. The planes dangled three layers thick.

"But, doctor . . ."

"There must be seventy-five airplanes in that room. You would not believe how much dust will collect on that many wings."

There was little she could do. The compulsion burgeoned in 1964 when John was in the second grade, after a family friend gave him a green plastic German Stuka as a Christmas present. The Judge saw immediately that this World War II dive bomber with its electric motor would be too complicated a project for his oldest son, and he helped the boy unbox his gift and spread the parts out on the dining-room table. Ordinarily, the Judge was good with his hands, a regular Mr. Fix-it around the house, a vigorous outdoorsman, a football referee for the South Eastern Conference, a forceful district court judge. Self-confident, capable. John watched his father fumble with the pieces. Mrs. Langford took a snapshot, which to this day captures the beginning of that disaster and disappointment.

"You boogered it up!" the Judge snorted. John watched his father become more and more agitated over the complications of rudders and winglets, glue and wires. The tiny parts slipped from his thick hands. "You boogered it up!" That was the last airplane model John and the Judge conspired to make. The boy decided he could do just as well by himself.

Unlike his father, who'd been a college football hero at Auburn and still gave a bullish impression, particularly to Atlanta lawyers, the boy had small bones and nimble fingers, much more like his mother's. He began to build ten-cent balsa models, twenty-five-cent rubber-band models, fifty-cent plastic jobbers. One summer during a vacation in the mountains of North Carolina, he collected them into an armada he called the Highlands Air Force, with HAF markings on each one, and launched them off the porch of their cabin. They flew impressively, except that some disappeared into trees. Soon John became a good tree climber, which, as a physical activity, was something of an appeasement to the Judge.

By the time he was in the fifth grade, he'd built every plastic model in existence and some, like the British bombers from World War II, which he especially liked, he made in duplicate. He borrowed against his allowance and played financial games with his mother to wheedle a few extra dollars. For a few months he switched to ships, but they weren't the same. It wasn't the process of building he enjoyed; it was the fact of flight. Not the fantasy. Because despite whatever anyone might think about John's activities upstairs in that four-bedroom Cape in the woods, he was attuned to the real world. His handicraft looked like toys, but they were only the adult world made miniature. He never read science fiction, like a lot of boys his age, because fantasies offered no real promise, only escape. He didn't watch television except for the news. When Soviet troops seized Prague in August 1968, he was building a rocket gantry out of Popsicle sticks in front of the TV set; listening to Walter Cronkite talk about the possibility of world war took his breath away. It was the inspiring moment, a turning point. By the time he found out that he could actually buy flyable model rockets ("space models," as the Europeans called them), he took to it with a patriotic purposefulness that went beyond fantasy. The aircraft were not a fiction to him; he was an explorer of the new frontier—zealous and dedicated—even though he was only eleven years old.

The national agenda clearly predicted that the country's destiny depended on space travel. A remark by the Russian father of modern rocketry, Konstantine Ziolkovsky, stuck in the little boy's mind: "The Earth is the cradle of mankind, but one cannot live forever in a cradle." John memorized inspiring words, just as he dwelled on the Cold War and questions of whether it would be the free world or the non-free world that would inevitably conquer space. NASA and Apollo and the new frontier provided the excitement of war—backyard blitzes—without the devastation of battle. He wanted a part of that. Progress and destiny and space travel. To John, model rockets were just the finest toys around. "As cool as grits," he'd say.

The first one arrived in a box for his eleventh birthday from Estes Industries in Penrose, Colorado. He'd run out to the mailbox every day after sending in his order. Then, one afternoon after band practice, he made that long trip up the driveway and there it was, the Estes Alpha starter kit—balsa and paper tubes, balsa fins, a red and white parachute, three solid-fuel rocket motors, and an electric launcher to fire it up. The Judge tried to put the rocket launcher together, but it got boogered again, and John took over. Once he'd assembled it, he disappeared to an open field near the house to ignite the Alpha, time after time after time. But the biggest thrill came not from the launches but simply from finding his order waiting in the mailbox. He'd made contact with people he never knew existed before—people like him.

A few months after the Alpha arrived, he took a trip with his parents to Colorado, where the Judge taught a summer course at the state university. John talked his father into driving to the Estes manufacturing plant in Penrose. It wasn't Cape Kennedy, but it didn't have to be. He was bug-eyed. It was the first time he'd ever been to a shop and had the privilege of pawing over the rockets before buying one.

Within fifteen months he'd built every kit in the Estes catalogue—eighty to one hundred rockets in all—and turned to making rockets from scratch. Following the Apollo missions closely in the newspaper and on television, John charted his own space rocket voyages in a series of notebooks, containing drawings, charts, timetables, and cost analyses for each. The table of contents listed fifty-eight rockets: Dove I, Sky 'Awk, Zoomer, Hustler, Bail Out!, Flaminarrow, SS Infinity, Pointy, Jupiter Excursion Module, Jetex II, Ballooner, Project Mouse Mile, and the Zion Missions. Project Mouse Mile, of course, was designed to launch a live mouse above the clouds. He carefully penned in timetables—a tiny filigreed script—for the launch, set for a midwinter afternoon, and held to a tight schedule.

Leave Home: 2:30
Arrive at launch site: 3:00
Set up pad: 3:05
Prep rocket: 3:15
Attach booster: 3:20
Check launcher: 3:25
Check power: 3:26
Load mouse: 3:29
60-second countdown
T-60 to T-50: check communication with track 1
T-50 to T-40: check communication with track 2
T-40 to T-30: final check
T-30 to T-25: 5-second power check
T-25 to T-20: check skies
T-20 to T-15: final check. At 15, 14, 13, 12, 11, 10, 9, 8,

battery sequence, the power is on 5, 4, 3, 2, mark and holding
0 Lift-off!
The mouse will be ejected at 13 seconds, driving down after
that, and recovery . . . ?

The idea of launching a mouse excited him because he'd discovered a way to ignite more than one engine at a time using the peanut bulbs from one of the Judge's old flash cameras. The Estes catalogue had some clustered engines, but usually when the first engine fired, the others short-circuited. He'd ruined a nice Saturn model that way. His own idea, which he considered the first really good idea he had ever had in his life, was to cluster a dozen or more engines and, using a small amount of electrical power to set off the flashbulbs, fire a series of wicks to engage the engines. He never lost any mice, either, though a few suffered bloody noses and one, which escaped its cage at home, slipped into his parents' bed one night, tickled his mother's feet, ignited a scream, and met its death when the Judge rolled over and crushed it.

By the time Neil Armstrong and Buzz Aldrin landed on the moon in July 1969, John was a fanatic. He'd memorized portions of speeches by President John Kennedy setting the course of the space program for the decade. He listened intently to a set of Time-Life recordings, hearing them as if the President were speaking only to him. "We set sail on this new sea," he'd repeat, dramatically, "because there is new knowledge to be gained, and new rights to be won, and they must be won and used for the progress of all people. . . . We choose to go to the moon. We choose to go to the moon in this decade and do the other things not because they are easy, but because they are hard. . . ."

Watching the first images of the moon landing late that night in July, he was mesmerized by the transmissions from Houston control to Apollo 11. His little brother, David, his sister, Ellen, and his parents all sat on the edge of the sofa. But John lay on his back on the floor, feet pitched up on the sofa, watching the images as they came in upside down from the moon. As if he were landing, too, listening attentively to Buzz Aldrin's final three-minute transmission. The rest of them went to bed, but he stayed up to listen, hovering precariously over the lunar surface, watching Walter Cronkite and the replay over and over again on CBS, all through the night.

Although it took him years before he could translate the jargon, he memorized these words, too: "Down two and a half. Forward, forward forty feet. Down two and a half. Two and a half down. Shadow. Four forward, four forward, drifting to the right a little. . . . Okay, contact light. Okay, engine stop. ACA out of detent, modes control both auto . . . descent engine . . . override off . . . engine arm off . . . 413 is in. . . ."

Who was this homunculus? Certainly not brilliant, but extraordinarily well organized. Slightly asthmatic and klutzy, but blessed with the gift of concentration and bullheadedness. Not particularly creative or rebellious, but full of desire to lead a new mission, to be a part of a great historical process. He walked fast, for a child, swinging his arms as if he were on a march. He talked fast, too, and if he was somewhat isolated from his peers because of his interests, he knew how to harness the energies of others toward his own ends. Though this slight-framed, blond-haired child with horn-rimmed glasses might have seemed misplaced among his generation, he fit well into an adult cadence. The ambition was clear. He wore his collars starched.

The latest in a line of children of privilege, John found inspiration in his grandmother, a woman who'd lived with considerable wealth, whose lineage could be traced to a nineteenth-century California robber baron, and who was a niece of Robert Woodruff—"Uncle Bob," as Nana called him—the man who built the Coca-Cola empire. His grandmother taught John the possibilities of the real world. She'd taken him out as a toddler to chase down cement mixers and search out tree grinders to learn how things work. She'd lectured him to spend his money on hardware rather than candy, to nurture a few projects, not to waste his time dawdling. It was Nana who would build a new house in the woods nearby—two large squares rotated one on top of the other—which looked like a star of David with Italian balconies. She wrote poetry and raised money for the Natural History Museum and flew around the world with Uncle Bob. Out to Wyoming to ride horses, down to south Georgia to shoot quail, off to New York for Broadway's entertainments, up to Cape Cod for the summer to relax at a family retreat called Greycourt. She taught John to dream as she had, to feel the natural sway of his own influence, just as he fell under the spell of hers. He built his ambitions on her spirit.

"As men have now landed on the moon, and we are rushing towards

Mars, I feel the time has come to move on to the stars," he wrote in his notebook that summer, using the nom de plume Ludwig Von Drake. "But first, as stepping stones, we must conquer our planets. Therefore, we must explore Mars and Venus, then Jupiter and Mercury. We are already pushing at Mars and I ask Congress to grant $4,000,000,000 (4 billion) to the space program for exploration of Venus, using Project Zion. Zion, after Venus, will help in exploration of Jupiter and Mercury. It will be a four-man capsule and a three-stage booster."

Within a few months, he'd organized a rocket club for Atlanta. He mailed out a dozen formal invitations to a select group of schoolmates and their parents, printed out an agenda for a meeting, and cajoled his mother into making refreshments. He cleaned the basement for an impressive assembly. The collection of twelve-year-olds that night struck the Judge and Mrs. Langford as not being quite the kind of children with whom they might have wanted John to associate. They seemed more like colleagues than little friends. They matched plaid pants and madras shirts, orange socks and black, low-top tennis shoes. One or two had Beatle haircuts. The rest looked like Harry Truman.

"They weren't your regular football types," Mrs. Langford said, herself a former beauty queen at the University of Georgia. "They were nerdy. No, not nerdy. You could say they were . . . interesting. It was the oddest assortment of people you ever saw. I remember one boy was simply huge. He was a great guy but totally unappreciated and disliked in school. A misfit. But he had a real adult sense of humor and was very intelligent. He played tuba in the band because he was the only one who could lift it."

After a brief invocation, John held up a magazine called *Model Rocketry* which had arrived recently in his mailbox. Published at the Massachusetts Institute of Technology, it contained news and photographs of rocket contests from around the world, technical data from tests of rocket launchers, flow visualization analyses, research and development reports, tips on using space-age materials, new designs that outclassed anything you could buy from Estes. The pages were chocked with altitude calculation formulas, sines and cosines, god-awful trigonometries that even John's math teacher hadn't been able to decode.

"This magazine," John said, waving it in the air, "is a real breakthrough."

Then, a serious explication, a studied exegesis, a page-by-page exposition.

The Judge's brow began to perspire because John was not moving the agenda forward. Looking around the room, he and Mrs. Langford saw the other parents, who were glad to have someone organize their peculiar children, smiling strangely and nodding vacantly.

Finally the Judge said, "John, I think we need to move on to the next point."

After the speeches, they went out and launched rockets, and subsequently, the Northside Rocket Club was born. John added his sister to the membership list to meet the quota for official certification, and then the editors of *Model Rocketry* placed the Atlanta section on their roster. In January, the Boston headquarters published a curiously uplifting note from the club leader: " 'The first rocket to fly in the new decade, January 1970 A.D. It flew in peace for all mankind.' This quotation was attached to a rocket flown by the Northside Rocket Club on January 1, 1970, at 12:00 a.m. Eastern Standard Time. The model, an Estes Alpha, has been retired by the section."

Later, the Judge and Mrs. Langford had a talk about John and agreed that he needed to diversify his interests. The Judge made a deal with his son. "Mild bribery," Mrs. Langford called it.

"If you earn your Eagle Scout award, I'll take you to the national rocket events," he said. "Son, you really will appreciate getting this Eagle later. You'll be glad you did."

Agreeably, John loaded up his fishing tackle box with glues and balsa parts and model rocket engines and struck out for Atlanta Stadium where, using conventional Scout teepee building techniques, he lashed together a few rocket platforms and launched rockets over the city. He took a course in public speaking and addressed audiences with patriotic fervor about the challenge of space travel. He ventured across Fulton County to fire rockets from industrial sites, from schoolyards, from the backyard of some of his grandmother's friends, such as Ivan Allen, the mayor of Atlanta. Quickly he earned one merit badge after another. He learned how to work this little system to his benefit. What was fire and smoke on the launching pad became statistical reports, which went into neat folders with fancy covers, just the kind of businesslike shebang that impressed men who sat on Scout councils.

The Judge took his son on scouting trips from the red-clay woods of

Georgia to the Minnesota wilderness. John earned merit badges in swimming and camping. After taking the Patrick Henry course in public speaking, John appeared on television hawking tickets for the Atlanta Scout Exhibition, smoothly pitching the event by calculating on camera the tangible benefits for buying the ticket ("You will pay only one-ten-thousandth of a cent to see every Scout who appears in the show!"). The Judge wanted the boy to expand his horizons, and encouraged him to be outgoing, intellectual, and physically strong. When John became a patrol leader, the Judge was pleased.

He earned his Eagle award by the time he was thirteen, and his father honored the promise. John had successfully bargained for traveling rights to regional and national rocket contests—from Seattle to Annapolis—and then to Cape Kennedy to watch the Apollo launches. Although the grip of airplane fever had been tempered by scouting, the trips to Florida sent him over the edge.

"Every American taxpayer should have a chance to see these things go," John would tell his friends in the Northside Rocket Club after a trip to the Cape. "It's a quasi-religious experience. It's like watching an atomic bomb go off. It's like an oasis of white light and this big needle on the horizon, pure white against a field of black, and when they get ready to launch the thing you hear the countdown over the loudspeakers and then nothing. You see a flicker of light at the base as the engines ignite and all of a sudden out of the side of the launching pad a huge blast of smoke and flames and in complete silence, like a dragon spewing flames horizontally, it comes out of the ground. It's like the sun coming up, so bright you can read by it . . . and you can't even see the rocket because it's followed by this incredibly bright tail and then a hundred feet up you see the plane in sight and it's still completely silent. But as it moves up you get hit by thunder. The sounds of ignition and explosions that get louder and louder and it shakes all the cavities in your body . . . it's just mind-boggling to watch it and watch it go and it's like it . . . like it embodies all of man's hopes and aspirations bundled up into this thing and riding off into the night."

Eventually, his father quit taking John on rocket trips. It was simply beyond the Judge's ken. His son was a goner.

* * *

"Who or what is a Guppy?"

John addressed his letter to *Model Rocketry* magazine on the third floor of Walker Memorial Hall at the campus of the Massachusetts Institute of Technology. There, he'd learned, the avant garde of space modeling plotted designs for weird boost gliders and tinkered with radio-controlled rockets, mixed explosives and advised companies like Estes on matters of pneumatic pressures and solid-fuel propellants. It was all in the magazine. In the model rocketry world, the guys at MIT were among the best. But their reputation rested as much on unorthodoxy as on scientific excellence. One of their members, Bob Parks, had a head of long blond curls and wrote a column for the magazine called *The Escape Tower*, which reported on his tests of untraditional designs—rockets with no fins, flying wings, rockets with wings that flopped open at the pinnacle of their ascent and glided to the ground by radio control. There was another fellow who wrote about controlling rocket trajectories through telepathy, and a third who had officially changed his name from Steve to Byrd. One member, who was known only as Guppy, appeared in the magazine wearing a top hat and tails, hauling extremely large rockets bearing loads that could destroy tree stumps. His name appeared often enough to pique John's interest: Guppy?

John would read the scientific articles in the magazine religiously, but he could not fathom this character in the top hat and tails, who was as tall as Abraham Lincoln and appeared to command a realm of the hobby given not so much to scientific truths as to astonishments. It was too mystical for John's sensibilities, even though he regularly saw that Guppy won awards in rocket contests around the country. When he wrote MIT and asked, "Who or what is a Guppy?" the response from the editors of *Model Rocketry* was formal and unsatisfying: "Like most good rocket builders, he's also a fine airplane craftsman. . . ." The MIT rocket builders nurtured an air of mystery.

He's obviously at the forefront of the hobby, John thought. Guppy was national-level; he was someone to watch.

To an eighth-grader in Atlanta, full of determination to escape his woods for the real world, these were the gods. He wrote them. He sent them roughcut ideas from his notebook. And eventually, he met them. In 1972, an MIT graduate student and former columnist of *Model Rocketry*, Tom Milkie, transferred to Georgia Tech in Atlanta and appeared one night at the Northside Rocket Club meeting. Milkie taught John new building techniques and introduced him to the kinds of analysis he needed to turn his childish models into aeronautical vessels. Milkie wanted to take John to the East Coast regional rocket contests in Virginia that year. And after sharing an evening meal of spaghetti with the Judge and Mrs. Langford—who needed to be reassured that Mr. Milkie was not a child molester but simply "a model rocketry type," as the Judge would say—he and John drove all night until they reached the contest site. They stumbled into a motel room where Bob Parks and Guppy sat prepping their rockets.

"I was awed beyond words," John said. "Because here were all these heroes of mine in the flesh, right out of MIT, with these incredibly cool rockets. They viewed me as sort of an urchin, I think, but an urchin with a calling card because Milkie was a buddy of theirs."

Little John Langford and his latest mentor, Tom Milkie, cleaned up at the awards contest while the giants of modern rocketry from MIT, Guppy and Parks, went back to the Institute empty-handed. From then on, using lessons he learned from Milkie and from the published works of MIT's avant garde, John collected awards and spread his name as a builder of scale models and originator of research and development projects. A reporter from the *Atlanta Constitution* drove out to the woods to interview him. A television producer called, wanting to shoot a video. John obliged them graciously. After a series of meetings, the producer handed John a script, which he read eagerly for the overdubs.

"My name is John Langford," he said, a slight southern accent blurring his quick and earnest speech. "I am sixteen years old. Go to Northside High School here in Atlanta." He'd developed a striding patter among adults, a glint of boastfulness, and a desire to display his talents.

The television camera caught him walking up the steps of the high school with the fishing tackle box in one hand, wearing a steel gray jumpsuit and aviator sunglasses, headed for the graveled roof three stories up to launch a rocket. During a misfire, his friends screamed, "It's your fault, John!" as the rocket dug itself into red clay hundreds of yards from the launch site. John's mouth turned up slightly—a hint of recognition—and he demanded quietly to the loudest of them, "Go get the rocket, Harold." They were children under the direction of a man-child.

"In the past two or three years I've begun to use my interest in rockets more and more in my schoolwork," he intoned. "The environmental studies project about temperature inversions is perhaps the best example of how I've combined the two."

The camera crew taped him on the rooftop and out in thatchy fields nearby, following him as he barked orders at a slew of friends who hustled to connect the electronic wires and set the gauges and punch the rocket controls. "You want to turn on the walkie-talkie, please?" he'd say. And someone would jump. Or, "Okay, stand back, we're ready to go," and his friends would repeat, "Okay, stand back, we're ready to go." The rockets blasted off, gushing gases that sounded like fire hydrants uncapped. Heh, heh, heh, his friends laughed. But John watched the model top out at twelve hundred feet and merely smiled as the parachute unfurled. He turned to the camera crew and said, "Now get that. That is beautiful. That will do. Definitely. Oh, that is good."

In 1974, when he was a junior in high school, the National Association of Rocketry selected John to represent the United States in the space-modeling world championships in Dubnica, Czechoslovakia. He joined a forty-two-year-old Ohio State professor, a retired colonel in the Army, and a handful of college undergraduates from around the United States to compete against full-time space modelers from Spain, Romania, Poland, Yugoslavia, England, Canada, and Bulgaria. His model, the Athena H-003, a 493-gram, 1:19 scale model, took six hundred hours to construct and served as his visa to advance, at last, behind the Iron Curtain. He joined the NATO forces against the Warsaw Pact, and as he would say, after "six seconds of fire, smoke, and triumph," he saw his name posted among the top ten modelers in the world.

He applied to MIT after the championships, writing the dean of admissions that he shared an interest in international diplomacy and aerospace. Most likely, he wrote, he wanted to help coordinate international space exploration. When he left for college the next summer, he was certain he would soon exchange his models for the real thing. In high school he'd experimented with film making and journalism, but nothing satisfied him like rockets and airplanes. At MIT, he could immerse himself. MIT would train him, mold him, prepare him to take the lead as an adult. At last he could get rid of these toys, the silly models. His notebooks were no longer filled with awkward sketches of model rockets, childish drawings of rockets shaped like flounders, lifted by helium balloons. They contained sketches of a life he'd shaped by design, in tandem with the American space program, plans for a life that would branch at any moment into industry, government, and the wizardry of the academic world. In tiny, careful script, he charted out the essential events of his childhood.

	SPRING	SUMMER	FALL
May 20 1957	I'm born!		Sputnik 1
1961			Nursery School
1962	Glenn orbits		Kindergarten
1963			1st grade
1968	Start rocketry	Colorado	6th grade
1969	Grandpa dies	Scout Camp Apollo 11	Broken arm
1970		Camp Michael	Patrol Leader
1971	Eagle	Ntl. Rocket Meet 13!	Freshman
1972	Canoetrip Dad's campaign	Ntl. Rocket Meet 14	First job, Baskin-Robbins
1973	Env. Studies Project #1	Europe trip	First film
1974	Quit job	Internats!	Senior Year starts
1975	Graduate	National Rocket Champs	Start MIT

2

August 29, 1975.

Sweltering station wagons laden with ambitious young freshmen nosed off the Mass Pike into Cambridge and unloaded thin, high-waisted little bodies misshapen under the weight of toaster ovens, stereo receivers, skis, knapsacks, ice skates, calculators, blankets, boots, deeply pocketed plastic baggies of marijuana, explosive SAT scores, and, most onerous of all, goals. Big goals. Enormous, looming goals.

They arrived, 1,155 in all, from 47 states, 30 countries; 234 from New York, 152 from Massachusetts, one from the Atlanta woods. They spilled out of their parents' wagons, peering at the grand entrance to the Massachusetts Institute of Technology, 77 Mass Ave, the Great Dome, which resembled the Pantheon of Rome but only—wait, this can be calculated—forty-two feet smaller in width. It was all sort of gray and not very pastoral, and really, even though the weather was warm and everyone acted excited to be there, the college was, in that special New England sense, an institute, after all. Cold and strange and terribly formal.

The Judge helped his son unload his luggage at the student center, until he could find a place to live, and then they said goodbye. At four-thirty, John walked alone over to the Great Court to join the freshman picnic and listen to President Jerome Wiesner's welcoming address. John had vowed to make a clean start, to make a break from modeling and focus entirely on his studies. He had left all his building materials in Atlanta. A thousand brainy kids made their own secret vows about relinquishing childish things and crowded around to hear Wiesner's speech, too. There in that grassy field the freshman class was honorifically greeted and rudely initiated.

"Hello—and a warm welcome to MIT on behalf of all of us—faculty, administration, and other students," Wiesner said. "Next weekend Mrs. Wiesner and I will have an opportunity to greet each of you personally, and hopefully your parents as well; meanwhile, please make yourselves at home and start being part of the MIT family. Beyond the pleasant task of welcoming you, I would like to. . . ."

Blah, blah, blah. The freshmen shifted and dawdled as Wiesner offered a statistical breakdown of the class. He discussed the challenge of technology, advised them on "habits of mind," and dallied over the usual academic drivel about learning how to learn. Then he started to warn them about something awful.

"The principal advice I give new students each year is a warning against becoming what's know locally as a Tech Tool," he said. "Instead of taking advantage of the many opportunities at MIT and in the Boston area generally, or getting a broad range of experiences and allowing all of your talents and the many sides of your personality to develop. . . ."

Lumped into this rhetorician's rhetoric he wanted to leave them a nugget of wisdom. But before John or anyone else could digest the president's message, a large banner unfurled behind the dais:

ABANDON HOPE, ALL YE WHO ENTER HERE!

Wiesner had been hacked. The freshman class had been hacked. The real right hand of warmth and welcome had been extended. Introductory niceties aside, the truth was the 'Tute wouldn't so much shape their fertile young minds as turn them loose in a culture of hard science, competition, and chaos. Some of them, truly, would not escape it. Some would become Tools, some would become Dweebs. They would learn to Flame and Gritch, they would learn IHTFP. MIT wouldn't cuddle or encourage them, and even though John thought he had been admitted into the breeding ground of astronauts, policymakers, and NASA engineers, he hadn't been accepted at all. He would have to learn a new language and new lessons. He was an urchin, like the rest, and he had just been blessed by some invisible upperclassmen who wanted their initiates to know the whole truth. MIT could be a strange and ungracious place. The sign, not the speech, made the most lasting impression of the day.

Overall, John found the Institute large and cold. The aeronautics department took very little interest in him. He wrote home; he complained. He felt the encroachment of a dormful of dorks. Not since summer camp had he been crossed by such a gang of maladjusted teenagers. But if at camp John had felt intimidated by tough kids who "smoked and could beat me up," MIT's crowd was full of sad creatures, and he was just as miserable.

The Judge wrote a letter to Sigma Alpha Epsilon, the athletic fraternity where he'd been a force during his college years, and tried to encourage them to welcome his son. SAE snubbed his spindly boy during rush week, and John wound up in MacGregor, known widely as "the second nerdliest dorm on campus." Within a week he wandered out into the rush orientation midway, where various fraternities and clubs had built booths to attract new members. There in the midst of all those signs and strange faces, he scanned for the Model Rocket Society.

When he caught sight of the booth, John was astonished to see who was there. It had been years since he'd seen these guys. Pictures in the magazine. "Guppy!" he said.

Hal Youngren, one of MIT's most accomplished rocket builders and an academic renegade, eyed him. He'd figured it would be just a matter of time before Langford showed up. Over the years, Hal, who was known to most people only as Guppy, watched students sweep in and out of the Institute like clouds. After six years at the university, sampling classes and labs from aeronautical to electrical to mechanical engineering without ever getting a degree, he had earned a vantage point that few students shared. He had developed a broad, keen view of the 'Tute, unencumbered by careerism and tunnel vision. "It's kind of like watching the weather," he'd say. "You look outside and there are all these clouds, but you don't realize that they come and die within fifteen minutes because there are always these clouds coming through. That's like watching people come through the Institute. There are always new people and every once in a while—puff!—something really interesting goes through."

With Langford, it would be hard to say how far his cloud would carry. From the moment he showed up at the midway, Gup thought, it was clear he'd join the club. John couldn't believe his old heroes were still at MIT—he'd assumed they would have graduated long ago and taken jobs in important places. Guppy saw how eager John was, too, how earnest and driven he'd become since they'd last met at a regional rocket event three years before in Virginia. The only question in Gup's mind was whether John would become a Tech Tool, as Wiesner said, or go ballistic like the rest of them. He has that high school science fair look about him, Gup thought. He looks like Mister Good Son America. It was true. Of all those eager young freshmen, John Langford was probably the least likely to abandon hope, despite warnings.

Within a few weeks, John found a proper home upstairs in Walker Memorial Hall, on the third floor next to the Student Homophile League and the Strategic Games Association, at the address of MIT's Model Rocket Society. Guppy showed him how to carve a wing using a razor plane and make streamlined wings that were aerodynamically superior to the plate wings he'd been accustomed to, and helped him cover them with Japanese tissue paper, which made his models as smooth as glass. Within a month, John wrapped wings almost as well as Gup. Enthusiastically, he volunteered to clean up the society office, which, like its members, was somewhat disorganized. He would meet them every afternoon at five o'clock, build models for a few hours, then go downstairs to eat. Since he was a freshman and, like most freshmen, had bought the meal plan with unlimited seconds, the veterans would follow him downstairs and use his ticket to refill their plates.

"Hey, guys," he'd ask, politely, "does anybody want something else to eat?"

If not, one of them would respond, "Our sufficiency has been surrencified. Any more would be obnoxious to our superfluidity."

"Oh," John would say. Then they would wander back to the club office.

It didn't take them long to realize how much they needed him. At the first official meeting in October they handed him a broom and elected him president.

"I was sucked in within a matter of days," John said.

The Institute's entrance at 77 Massachusetts Avenue was a model of classical Greek architectural styling, a bold gray facade sustained by impressive columns. From outside, the buildings suggested that MIT was a place of natural order and elegance. That, at least, matched its reputation. But inside, under its sky-high dome, was a hub, which spun off a colorless network of floors like the center of a maze. They shifted out into long, pale interlocking corridors and underground passageways linking building to building so that no one ever had to step outside to traverse the 128-acre campus. Compact and confusing to navigate, the Institute's geography was further obscured by a system that assigned each building a number, as opposed to a name, and scrambled their sequence so entering and exiting a particular hallway could be a bewildering experience. Building 9, for example, housing the center for advanced engineering study, hinged directly to Buildings 33, 13, and 7. To go from the Great Dome to the Hayden Memorial Library, one walked straight down a quarter-mile hallway—known as "The Infinite Corridor"—through Buildings 3, 10, 4, 8, 6, 2, and 14. A guide to the campus advised that it would be useful to know that the main campus east of the Great Dome had even-numbered buildings and to the west buildings had odd numbers. Thus, buildings west of Massachusetts Avenue were designated with a W, those to the north of the Conrail tracks were assigned an N, those east of Ames Street took an E, and buildings west of Massachusetts Avenue but north of the railroad were called NW. Problem was, the numbers remained the same. Consequently, W-15 was the chapel, but NW-15 was the magnet lab. Since some buildings had additions, a few places, such as Building 17, segued to an addition with the same number (17A, in this case). But these were not to be confused with like-numbered buildings with letters, such as 17E, which was on the other side of Ames Street. The advice given to newcomers was to try walking around and getting lost until the scenery became familiar.

For most people, the system was a frustrating and unnerving mess, a visible paradigm of MIT's occasionally muddled, high-minded theoretical approach to engineering sciences. The geography, in essence, mirrored the experience of being there. But there existed a whole subculture of students who used this insight primarily to wreak havoc around the buildings and conquer the campus. Those students whose spirits couldn't be contained by 'Tute's occasionally limited vision, or deterred by the academic and geographic hurdles heaped in their paths, had come to be known as hackers. They didn't lose themselves in bookish marginalia.

Compared with the Tech Tools—the nerds (a.k.a. Dweebs) who'd given MIT a reputation as a haven for data-munching bookworms—hackers, at their best, rebelled through the act of creation, and put radical engineering on a par with avant-garde art. Instead of conforming to the MIT academic environment and regimen, they attempted to master it.

Hacks took many forms, the best of which could be characterized by a sense of mischievousness, craftsmanship, adventure, creativity, and scientific excellence. At the simplest level, hackers had made a game of getting around MIT's labyrinthine system of interior pathways and corridors by traipsing rooftops and slipping through steam tunnels to explore the campus. The challenge was to sneak inside hard-to-reach quarters, such as the Great Dome, the abandoned elevator shaft in Building 56, the bricked-in shower in Building 14, the false ceiling in Building 1. For some, the game of exploring was made more interesting after smoking a few joints, when the drug played tricks on the senses and challenged hackers' mental faculties and imaginations most.

"Then it was like you couldn't even remember where you were and it was like being in some kind of spaceship," said one of the rocket society members from the mid-seventies. "It's all inside and so artificial that it can be really spooky. There are these huge halls and under these huge halls are these other huge halls and under them are these basements and then steam tunnels and connecting rooms and passages. It's like some big Dungeons and Dragons game. And it has this background sound everywhere, this artificial humming.

"What's weird is there are some places you can't get to. Like you go up from a basement and then you get to a door and it's locked. So you look through the little window in the door and you've got to get across a little corridor to another door. So you say, how do we get there? It turns out that you've got to go way the hell out of the way, out of the building up this way, and like half an hour later you get into the corridor and you're behind this door and it's locked. So you're only twelve feet away from where you started and it's taken you half an hour to get to where you are."

If a team of hackers had mastered the fine craft of locksmithing—as the best did, in time—they carried the adventure beyond simple exploration. After jiggering a lock to an inaccessible hideaway, they would leave behind some evidence of their cunning.

Once a group of hackers wired a working phone booth on top of the Great Dome, and when university officials went to examine the phone, it rang. A smiley face was plastered over the radome on the earth sciences building one year. A group of hackers mounted a twelve-foot, twin-threaded, left-handed screw made of wire mesh and papier-mâché atop the Dome once, where it remained until university officials decided it was not in good taste and unscrewed it. Another time, an enormous fiberglass cow, taken from the Hilltop Restaurant (a steakhouse in nearby Saugus), wound up on the roof of the Great Dome. Hackers even attempted to attach a gigantic plastic nipple to the top of the electrical engineering dome, with a sign saying MAMA MAXIMA one year, but campus police caught them in the act.

In other years, hackers made more literal statements, like the warning unfurled at John Langford's freshman picnic in 1975. During another student orientation week, they left a sign on the electrical engineering building saying: "Welcome to Havahd, Hackito Ergo Sup." The word "concubitus"—which one professor declared "a Latin word which our lads are too erudite to perform"—was scrawled in enormous letters atop a building facade. Traffic signs saying: "Slow, children playing," had been changed to: "Slow, Gnurds." A professor in Building 4 came to work one day to find his nameplate changed to read "Department of Alchemy." In the biology department, Building 16, a laboratory door was, for years, designated as the office of Prof. R. Catesbeiana, a man whose name never appeared in the faculty listings or phone directory but whose opinions could, at times, be found in the letters to the editor section of *The Boston Globe*, rambling on angrily about the U.S. bombing of an Icelandic volcano or the loss of an obscure Boston building, such as "the Plywood Palace." Prof. R. Catesbeiana, it was discovered, also had a Social Security card, but he was in truth a hacker whose nom de plume was an American bullfrog's Latinate *Rana catesbeiana*. And, of course, perhaps the most omnipresent hacking sign was IHTFP, which appeared on domes and T-shirts and at campuswide events for years, expressing the commonly shared attitude among hackers for dear old 'Tute—I Hate This Fucking Place.

The best hacks, though, were the ones that either required scientific expertise—enough engineering excellence to confound the school's administrators and bureaucrats—or were so imaginative that they transcended accepted scientific wisdom to become legend.

Legendary among hacking tales, for instance, were the imaginary measurements that had become part of the Institute's tables of metrics and customary equivalents. For example, there were the lines along the Harvard

Bridge, repainted every year since 1958, which designated the span in terms of an MIT measurement called "a Smoot." More than twenty years ago one Oliver Reed Smoot, Jr., was wrapped in a blanket, taken out to the bridge by a group of hackers, stretched out lengthwise, and rolled end to end from the Harvard side of the bridge to the MIT side, with chalk marks and paint dabbed until they measured 364.4 Smoots. To this day, it's said, Boston police officers jot down the nearest Smoot number when filling out an accident report on the bridge. Similarly, in 1972, residents of Baker House spent four weeks planning a hack to drop a 600-pound piano off the roof of a six-story building on campus. Using a high-speed camera to document the event, the idea was to measure the volume of a dent created by the impact of the piano, there-to-fore to be designated a Bruno (named after Baker resident Charlie Bruno, who master-minded the plan). On October 28 the piano, covered with IHTFP slogans, was dropped and the Bruno inaugurated.

Child's play, of course. Not the type of behavior that would gain John Langford's approval. Yet almost from the moment he arrived he became president of an organization that included some of the most notorious hackers the department of aeronautics and astronautics had ever known. They were extraordinary craftsmen, inventive engineers, obsessed with the dream—and science—of flight. But they did not like to sit in classrooms or write scholastic papers or complete problem sets. They liked to fly. They liked to have fun.

Hyong Bang, for instance, was a brilliant aerodynamicist who pushed his knowledge so far in the lab that he nearly toppled it. Once, while exploring the netherworld of solid-rocket propellants, he suffered a problem with microprocity and, instead of burning, the fuel detonated. Shock waves inside the steel chamber ruptured the internal diaphragm that was supposed to relieve the pressure from explosions, ripped the steel door off its hinges, and embedded it in a row of lab shelves. The explosion dented the roof of the chamber, shattered the shatter-proof glass in the door, startled everyone in the building, and effectively banned Model Rocket Society members from the lab forever. To make sure the rocket society stayed out of the mechanical engineering department, a professor with a forklift showed up shortly afterward in the basement of the aeronautics department, hauling the tattered chamber. "I'm through with these guys," he told one of the aeronautics instructors. "I've had it." And he left the steel box in Building 17, where it sat for the next ten years.

John also found himself sole representative of a few guys who organized the so-called commando group, a paramilitary band of long-haired rocket hackers who wore fatigues, carried rifles and grenades, and spent their weekends in mock combat with the campus ROTC at regional Army bases. As members of the rocket society, they could requisition chemicals to make grenades, bazookas, and booby traps. "The per capita ammunition expenditures from the rocket society were three to four times higher than the ROTC," said one former rocket society member. "We had M-16s with blanks and made our own explosives, and filled rockets up with a moderate amount of ammonium perchlorate and aluminum—standard rocket stuff—to make bazookas. We were doing the explosives and getting the stuff from our lab supply account. At one point our faculty adviser, who was not too closely involved with the club, sent us a letter saying he'd seen all these requisitions and he figured they had no other utility than to make explosives, and ordered us to stop."

Once, a couple of them used their commando booby trap to bomb a young man who hung out at the campus cafeteria, a peculiar student who, they noticed, would not touch doors with his hands and would only use envelopes of sugar that had pictures on them. He had come to be known as "The Tech Man." They set their trap late one night along a walkway leading from the student center library, a popular hangout for despised Tech Tools. They watched him from the roof of a nearby dorm as he shuffled past. Nothing happened. When they went down to check the explosive, it suddenly detonated and one of the rocketmen spent the evening at the hospital emergency room having his charred groin tended.

Such were John's charges. But they were also the best hands-on engineers in the department. While many undergraduates came into the lab thinking a woodscrew was made of wood and many professors wouldn't think of sullying themselves over a milling machine, the modelers lived to build. Unrecognized by the faculty and their peers, they'd become expert at making two-ounce rockets of balsa, paper, and lightweight space-age materials, including exotic glues that were used on the Space Shuttle. They sized their miniature machines to fly at three hundred miles an hour, straight, true, and thoroughly tested. With ordinary Q-tips, X-Acto razors, clothespins, alkaline batteries, cotton twine, tweezers, needle-nosed pliers, sandpaper, and scissors, they created incredible "kludges" (MIT lingo for any mechanical contrivance that appears unlikely to work but does anyhow). Day and night, they computed thrust and drag and pitch and yaw, studied the boundary layers of their fingernail-sized fins, toyed inside MIT's low-speed, low-turbulence wind tunnel, developed new instruments to measure motion and forces on model rockets. Then they decided that most model rocket designs were too conservative, after all, and that they could get more performance with even lighter-weight materials, that they could design gliders to attach to their two-ounce, 300-mile-per-hour rockets, and dress them up with lightweight electronics and radio-control equipment to emulate a Space Shuttle. They built series-staged motors and static test stands to perfect multistaged rockets, calculated the optimal size for parachutes and recovery streamers, and made extraordinarily accurate tracking devices, called theodolites, to measure the elevation and azimuth angles of their rockets. They built models, mostly, and the best of their crew avoided classrooms, but worked in a very technical tinkerer's paradise on the outskirts of academia, unlikely ever to graduate, unlikely ever to leave.

In time, John became one of them, the one who would find them money and handle the logistics of getting them to model rocket competitions and make sure they didn't go ballistic all the time. Going ballistic: a rocket society term, meaning soaring out of control; blowing up a lab; forgetting to sleep for three days because you're working on a rocket glider; lobbing rockets stuffed with explosive blocks of sodium into the Charles River at night so the water appears to burn, shimmer, smoke, and crackle; firing rockets with explosive payloads from the third floor of Walker Memorial into the midnight skies of downtown Boston, watching the fireworks display, and listening to the city roil with the sound of puzzled police sirens. When rocket hackers went ballistic, the skies over Boston lit up.

During his first semester, John had to drop one class and muddled through his others. MIT was not what he'd expected. He'd entered college thinking he'd find a large community of people like Guppy and Parks and Tom Milkie. What he hadn't realized was they were the renegades of the aero department. They had their own vocabulary and own definitions of aeronautics education. They didn't accept the Institute's stated requirements as synonymous with learning. "Building airplanes and rockets is completely inappropriate behavior around here," he told his family. "From what I can tell, the faculty doesn't care about making you a good aeronautical engineer. They just want good material for graduate school."

Unlike some of his friends, John kept going to class, completed his problem sets, and made decent grades. In the classroom he learned calculus, physics, chemistry, and political science. But he learned aeronautics from Gup and Bob Parks.

Articulate, clean-cut, and politically astute, John dealt with the bureaucracy at MIT and for the next four years, from the third floor of Walker Memorial, he worked to help his friends win all the major model rocket awards. They believed, deep down, they were the best rocket makers in the world despite their academic standing at MIT, and John made certain all systems were go. With the kind of fervor more often found at the foot of an altar than the base of a launching pad, John became, in some small but significant way, the manager and goodwill emissary of the MIT rocket team. Their confluence of talents was as ironic as it was successful.

"Given deliberation," Guppy said, "John would always come out and do the right thing. He rarely went ballistic."

3

After three years at the 'Tute, John had breezed the basic aero courses and impressed faculty as stalwart MIT material, able to work in industry, government, or academia. By January 1978 he'd accepted an offer to work at Jimmy Carter's White House as an intern in the Office of Science and Technology Policy, assigned to an analysis of the nation's aeronautics policy. He had his own office, his own WATTS line, and entrée into the NASA establishment. "Life," he wrote in his college journal, "has begun to move so fast."

Ironically, as his options multiplied, John remained bound to the Model Rocket Society. A childhood dream of becoming an astronaut tempted him. He considered a career in politics. For his girlfriend's sake, he wanted to settle down after college into a reasonable nine-to-five job, but for his own sake, he wanted to explore, to build, to cash in on the promises President Kennedy had made at the beginning of the previous decade. (At school when the dorms had stereo wars, and the neighboring Tools blasted MacGregor Dorm with the Rolling Stones, John and his roommate fought back with the old Time-Life recordings of Kennedy's speeches from the early sixties).

As he walked up the large, open stairwells of the Executive Office Building to his desk on the third floor, John mulled over a future increasingly muddled by unsatisfying options and childhood dreams. He really preferred flying rockets with Bob Parks and Guppy Youngren, competing around the world, building their own models. He derided himself for even considering the idea, and promised that he'd give it up after the world space-modeling competition in Bulgaria that fall.

Still, in the midst of his life-planning dilemmas and important work at the White House, John spent all his spare time on a scale model to take to Bulgaria. He slipped into the National Air and Space Museum during lunch hours and broke loose a secret set of technical data on the X-2, Parky's scale-model entry into the world event. At heart, he was a radical rocketman, too. A conflicted aeronautics hacker, but a hacker nonetheless.

Unlike a generation of young computer hackers across the country, whose energy and talents invigorated a new industry, those obsessed with the dream of flight could hardly expect to channel their vision into professions that needed entrepreneurs or innovators. Despite President Kennedy's promises of their childhoods about space and the new frontier, there was no demand for people who wanted to build, design, and fly rockets or airplanes. There were plenty of slots for John and Guppy and Parks in one job or another, but no pigeonhole could contain their imagination and ambitions. During the space age, they realized, an MIT education could do little more than reserve a place for them among a fleet of aeronautical bureaucrats.

It was a significant change for MIT, too, this dissipation of the dream. The department of aeronautics and astronautics had once reveled in an engineering aesthetic that not only linked craftsmanship with science but placed its graduates squarely on the edge of a new frontier. Some of the first courses ever taught in aeronautics had started at MIT one hundred years ago, when students at the Institute built wind tunnels to test air pressure around surfaces shaped like dinner plates and birds' wings. In 1896, a thesis by A. J. Wells described the construction of such a tunnel—30 inches square with an airspeed of 1,300 feet per minute—and the results of his tests are considered the first documentation of the importance of airflow on the upper surface of a wing, a pioneering achievement. The Wright Brothers' propellers were tested at MIT using scale models during those years. In the winter, students experimented with gliders up and down the Charles River when the water iced over, skating with kites or towing improbable aeroplanes fitted with pendulums, fired by compressed air pistons. During the twenties and thirties students and professors spent thousands of dollars analyzing the flight of sea gulls along the Charles. They lured the birds with fish, captured, killed, and mounted them in wind tunnels for testing. For decades they set out on a search for the secrets of pure flight.

During the thirties, forties, and fifties, young men who came to MIT learned many of those secrets and went on to shape the airline industry or to lead the nascent space program. J. S. McDonnell, Jr., who became president of McDonnell Aircraft, graduated from MIT in 1925. James H. Doolittle set transcontinental speed records in airplanes during the thirties and became a legendary World War II flying ace. Robert Seamans, who served as deputy administrator of NASA during the Apollo program and later as Secretary of the Air Force, labored around the Wright Brothers Wind Tunnel in Building 33 during the forties. James Stark Draper, who tinkered and taught at the Institute for half a century, pioneered high-altitude flights for taking weather data during the twenties; built gyroscopes during World War II to compute the lead time needed for shooting down dive bombers and kamikazes; developed inertial guidance systems for unpiloted planes during the fifties; brainstormed a navigation system that guided men to the moon in the sixties. Buzz Aldrin's father, Edwin E. Aldrin, took a master's degree at MIT in aeronautics in 1917, and years later Buzz earned a doctorate; in 1969, Buzz joined Neil Armstrong for the first walk on the moon.

During the golden age of aeronautics MIT professors worked with slide rules, not computers, doodled with 6H pencils, built designs they could test in the Wright Brothers' tunnel, and seemed as comfortable at work in their shirtsleeves or overalls as in coats and ties. Twenty years ago, professors from the department would leave their offices to share lunch at a cafeteria across the street from Building 33 on Mass Ave, where they'd argue out ideas in pencil on marble tabletops and, once done, swipe the markings clean with a damp napkin and head back to the wind tunnel to test their theories. It was a time when the most learned of them still knew their way around all the steel and plywood and aluminum yards in Cambridge. "It was a great place to build things," remembered Joe Bignell, an MIT professor and director of the Wright Brothers Wind Tunnel from 1939 to 1969. "There was a spirit of lighthearted, competent application."

Over time, though, the school increasingly became a place for scientific specialists, professors who weren't even interested in building "paper" airplanes and students who would fashion themselves in the image of theoreticians, not the creative engineers of previous decades. For those still enamored of the dream of flight, as the new generation of radical rocketmen were, adulthood offered an uncomfortable dichotomy between immersion in what most of the world would see as a childish fancy and a career track that would place them, whether in government or industry or the academy, amid an inglorious sea of gray desks.

For youngsters who entered MIT captivated by missions into space during the sixties, the 'Tute could be a grossly disappointing place. Some accepted the fact that bureaucracy had displaced adventure; they were the Tools. Some, like John, struggled to reconcile themselves to the unsatisfying choices and maintained some hope that the promise hadn't been a lie. Others—the hackers—entered their own reality, living according to standards they alone set and voices they alone heard.

While John charted the future, his friends in the rocket society, Bob Parks and Guppy Youngren, entered their tenth year as undergraduates—off and on—at the Institute. Impervious to the ironic dictates of the space age, they ignored the scholastic bureaucracy, signed up for courses and abandoned them at will, regularly forgot to turn in homework, and used what they learned to improve their craftsmanship and advance their knowledge of computers, electronics, and aerodynamics. Guppy and Parky admittedly had perhaps the highest standard deviation of grades—a spread of As and Fs—in the history of aero-astro, and it appeared unlikely that they would graduate any time soon. They didn't chart out a life plan, like John. While he fretted, they just kept building rockets and weird new devices that looked and worked like the Space Shuttle in miniature. They called them radio-controlled boost gliders.

Harold T. Youngren changed the history of the Ford Motor Company at a time when the company's greatest weakness, aside from its financial doldrums, was its engineering. A central figure in the development of the automatic transmission at Oldsmobile and Borg-Warner, Youngren catapulted into management early in his career, but found himself dissatisfied in environments removed from hands-on engineering. His switch from an executive position at Borg-Warner to chief engineer at Ford during the forties marked the rebirth of the automobile company under the direction of Henry Ford II. Youngren's engineering ability exceeded anything Ford employees had seen in a manager and his craftsmanship was remarkable. It's said that in 1947, not long after Youngren arrived at Ford, an aged Henry Ford and his wife, Clara, visited the laboratory where Youngren and his team were building a mock-up of the 1949 Ford. Clara broke the door handle of the clay model when she mistook it for the finished product. "It looked so real," she said.

Guppy never really knew his father, who was an old man when he was born, because business frequently kept him away from home. But the child inherited certain remarkable characteristics. From the time he was a toddler, growing up in a wooded section of Annapolis, Maryland, he always hauled around a fistful of screwdrivers. "And you couldn't take them away from him, either," Mrs. Youngren said. "I was always afraid of his falling and sticking himself in the eye, but instead, when I went to take them away, I was the one who got the black eye." He called them "drab-drabbers" and he used them to take apart anything he could and to pry open those things that made him curious. When he didn't fiddle with screwdrivers, he read; when he wasn't reading, he was observing—lying on his stomach tracing the pathways of ants, eyeing the stars from a telescope. "Life was never dull for him as a child," his mother said. "The world was a big, wonderful place." His eyes were lit by a startlingly blue sparkle.

By the third grade, he'd read Tom Swift, shared the adventures of Rusty Spaceship, gleaned what he needed from *The Golden Book Encyclopedia*, and decided he could tap into all that mystery by himself. Puzzling over vagaries of flight and fiction, he made a few judgments and went into his dad's shop to test his guesswork. By the time he entered grade school, he had designed and built toy submarines, jets, and model airplanes.

"There was always this intense feeling for making new things," he said. "Things I'd heard about or read about in a book. Making things happen. Trying different materials. Different compounds. I mean, you sort of learn the rules. All of life, all of engineering, is basically taking a set of tools and materials and learning those tools and materials until you can start playing the game. And then you start making new games to play. That's the way it is with mathematics, for example. You learn the rules of calculus, and all these things suddenly become possible that were never possible before."

His education began with a $500 set of Time-Life science books his parents bought him, which taught the basics of chemistry and physics and introduced him to elementary experiments, the secrets of hydroponics, and germination. By the time he learned to synthesize chemicals in the sixth grade, he knew enough to perform experiments, lecture, and teach his classmates at school. The problem—a mixed blessing, really—was that he couldn't be contained by a classroom. He performed well in school, but he had less use for it than he did with the out-of-doors, where nature played experiments all around.

Guppy, like John, was also enthralled with airplanes and rockets. He decorated his bedroom wall with pictures of the X-15, the Jupiter and Mercury space vehicles. He coated his ceiling with a map of the solar system. He spent a lot of time learning star clusters, watching galaxies, and viewing nebula from a telescope at home. He wanted to know how real rockets worked; he wrote to NASA and sought technical reports.

"*Britannica* had this thing where you could send away with coupons that came with your *Britannica* series," he said, "and they'd send you a report from a place—not like a think tank but a research organization. And they'd send you a report on anything you happened to be interested in. The crazy thing about it was the people who did this had no limit on what they would write about. They didn't edit it at all. So if you said, gee I'd like to know how to make explosives or rocket fuel, they would research this and send you all the relevant stuff they could possibly find.

"Basically, I was interested in kind of a combination of things. First, rocket fuel, simple rocket fuel . . . I mean, how do you make gunpowder when you're thirteen years old? Rocket powder is something you read about in your history books, and you're not quite sure what goes into it. I mean, you know what goes into it but you don't know what works best. And it goes on from there. From your knowledge of chemistry you say, well, there's an oxidizer and a fuel. And in this case, the fuel is sulfur and carbon and the oxidizer is potassium nitrate, saltpeter. So you say to yourself, golly, I know there are more exotic oxidizers than potassium nitrate. Look over here. And you look in your gigantic science book and you see—aha!—perchlorates! Perchlorates are much more unstable! Perchlorates must be much better. So then you try it out and, yep, perchlorates are much more unstable."

He singed his eyebrows in explosions. He obliterated a tree stump. A classmate who obtained one of his little bombs accidentally set off an explosion in the school cafeteria one day and was handily expelled. Silly stuff, but it lead Guppy further afield.

"And then I was interested in fireworks," he said. "I always loved fireworks because they were so ephemeral and they were light. It's hard to describe exactly, but you know how butterflies seem so much more beautiful because they don't last very long? Fireworks are the same kind of thing. It's something you see once and you enjoy it for the sheer uniqueness and joy of it during that instant. They're also much more exotic in a way than simple explosives and much harder to make. It's not even almost a science. You go back to the Dark Ages. People don't talk about it, they don't even write it down. Figuring out how to make the different colors can be tough. . . . Copper compounds and barium compounds and you have to add the right amounts and the right kinds to make these different effects.

"So early on, those two things—fireworks and explosives—were the pyrotechnic part of the puzzle for me. Part of my education was fire and smoke. The other part came later."

To hear him recount childhood experiments that took place in his backyard is like entering the charmed territory of Tom Swift. What Gup says about his early favorite fictional character was precisely how friends came to think of him: "He fills your head up with all kinds of goofball ideas, gadgets, and weird things—this kid with enormous industry who can do anything he wants, and build and play all day (actually work all day), and it would never take him very long to develop anything because it was just a fantasy and he never grew old."

Gup stirred tangy compounds. He launched mice and performed fireworks, fooled around with ESP and telepathy and hypnotized his friends. Very early he stumbled into the guts of calculus by making his mother stay up late with him writing out long sets of algebraic equations by hand as a way of plotting his rockets' trajectories. A fascination with photography took him on explorations of the night skies and into the micro-realm of cellular biology. His fingers were practically unusable at times because he'd jab pins into them to draw blood cells to photograph. It wasn't just rockets and airplanes he was drawn to; it was nearly everything mysterious and, by science, manageable.

Above all else, though, Gup wanted to fly. For a child at that time, model rocketry was the only way. The models were pretty small, and they didn't make much of an impression, but they were as close to exploring the solar system as he could come. He joined the Annapolis Rocket Association and all his scientific, creative, and pyrotechnic interests coalesced into a little footlong plastic wiener of aerodynamic art. He saw them first at a science fair demonstration in Annapolis. The trick wasn't just to launch a mouse or get the rocket off the ground. The goal was to win national awards in flight duration, speed, altitude, and glides. The metier combined calculus, physics, chemistry, and basic workbench argot. The result—dazzling displays of fire and smoke. For Gup, the little sky bullets offered significant possibilities.

"You'd look at them go and say, is that it? What's the big fuss about? But there was a lot behind it. Mostly you didn't want them very big because you didn't want them very draggy. The more weight they had, the more drag they had. So most of them were small. That's the first thing you'd think. The second thing you'd think is, they go awfully damn high. So you say, boy, what made it do so well? And you find out it's because somebody spent a lot of time making the fins the right size, and the body smooth, and the nose cone the right shape, and matching the weight to its engine because it turns out there's something called the ballistic coefficient. A ballistic coefficient is a function of how much drag it has and how much weight it has.

"That stuff is real important for rockets because a model rocket fires for only a second or a split second and then coasts. The ballistic coefficient is sort of a function of the weight and the drag. The higher the drag, the less coast you get, unless it's heavy. If it's heavy, it can't go that fast while it's boosting the rocket engine. It turns out there's an optimum. If it's too light it goes sshhwwht! and then stops. If it's heavy, it doesn't go very fast but it coasts a long way. Somewhere in the middle the rocket goes the highest. And there's a trick to that—the trick is understanding the math or the charts and formulas and calculations. And that kind of stuff had just hit model rocketry when I got involved."

He used the library at the Naval Academy to track down NASA reports and master's theses related to model rocketry. Excitement over the Apollo program spurred graduate students into active experimentation with model rockets, and the data became a prize for young hobbyists like Gup. Space research at NASA trickled down to the playground. Gup caught up with an engineer at the Goddard Space Flight Center who'd written his master's thesis on sounding rockets, essentially big model rockets that were fired into the ionosphere and traveled through the closer reaches of outer space carrying instrument tables. Sometimes they'd be fired through a hydrogen bomb explosion. Sometimes they'd carry photographic equipment and give glimpses of the sun. But for a hobbyist, like Gup, who wanted to know the precise way to calculate the size and shape of the fins for his models, the rockets provided excellent prototypes; NASA's research reports explained everything. "Once I got my hands on his technical reports," Gup said, "I was set."

The Model Rocket Society at MIT also provided excellent tips. "There was this book called *The Fundamentals of Model Rocketry*, published by the MIT Rocket Society," he said, "and this book was absolutely amazing. It was full of altitude trajectory calculations, explanations about drag, dynamic stability. Exactly what you use in a real-live engineering environment where you'd build fighter planes or rockets. So at MIT there was a collateral kind of development between model rockets and what people were learning through the development of the space program. For people like me it was just really fun. It was current. It was forward-thinking."

The rocket association Gup joined as a teenager eventually merged in his imagination with the Tolkien Society of Annapolis, of which he was also a member, and the results startled model rocketry association members up and down the East Coast. It was, Gup recalled, a time of invention but little practicality. "It was tough to get it through my head that you didn't want the ultimate rocket necessarily," he said. "You wanted something that was good enough to push performance as far as your skill and desire and materials could stand, but just a little bit short of being too far out to be practical. And of course at that time I just hadn't grasped the point. The point was reliability, and testing your ideas time after time after time, until you win. That's how the world really works."

Early issues of *Model Rocketry* magazine, 1968–70, featured many photographs of a tall, lanky, long-haired boy, only identified as "Guppy," testing the most incredible flying objects at the most unlikely places. He's shown heaving a five-foot glider called Gargoyle off the fire escape of an

Army barracks at 1:20 a.m. in Harrisburg, Pennsylvania, surrounded by a group of gaping youngsters. There's another picture of him wearing a black top hat and black turtleneck sweater, delicately displaying a featherweight, footlong glider in front of an Army helicopter. Another photograph shows him in a motel room, stretched out on a bed strewn with rocket-building paraphernalia, working on a wing. In another, he's seen in top hat and cape at a launch site prepared for the Gargoyle's liftoff, while a stunned rocketry official stares, mouth flopped open, scratching his head, agog.

In May 1970, during Gup's last year of high school, he happened to register for a series of contests at the Indiantown Gap Military Reservation, a weekend event set aside exclusively for testing radical designs. He arrived from Annapolis with his black top hat and cape carrying the Gargoyle and a slew of Tolkienesque notions, obviously closer to the nearer reaches of too far than anyone else in attendance. Except for one other. Gup met a curly-haired MIT freshman who'd driven down from Boston with a wondrous haul—a radio-controlled boost glider with an eight-foot launch tower, a rocket with an engine nearly sixteen times the allowable thrust, and a wing animated by twenty-seven moving parts.

Jesus, boy, that looks like more fun than I've ever had, Guppy thought. It's a glider and it goes up like a rocket and it's radio-controlled. I've never seen anything like it. But then, Bob Parks had never seen anything like Gup and the Gargoyle, either.

After three days, during which the modelers lost rockets in a mock Vietnamese village the Army used for jet-strafing runs and rushed events between the movement of tear-gas clouds, which drifted from the military reservation onto their launch sites, the two boys had, for the first time, met their match. Any device one could imagine, the other would build. Every day they arrived at the launch sites to fly a new creation; a seventeen-foot-long rocket made of paper that required six people to move it, but that could fly fifty feet before buckling; a boost glider built to look like a witch on a broom, with brass door hinges actuating the wings and dowels supporting the wings' leading edges; a flop-wing rocket made of five-minute epoxy and spare parts rounded up in a motel room. They set records and amazed the judges.

"Normally, I do not make friends easily," Parky recalled. "But that one was instantaneous. The top hat and tails to me would have ordinarily been a major discouragement. But he was doing some interesting things. Aerodynamics is weird—a lot of things people feel by gut are just wrong. And then there are a lot of people who can make something work but they don't know why. Guppy wasn't like them at all. He knew exactly what he was doing."

Once Gup showed up at MIT as a freshman in 1970, he and Parky became roommates and critical foils for each other's innovations. Reliability mattered more, perfection less. Even when Parky's education was stalled by a four-year tour of duty in New York, California, and the Marshall Islands with the Coast Guard, they continued a lively correspondence by phone and by letters, honing ideas for lighter, faster, more efficient radio-controlled boost gliders. Guppy would phone Parky with the results of his experiments and Parky would send Gup photographs of new designs he'd tested. Radio-controlled rockets became an eight-year passion, uninterrupted by school and military service.

Dear Fish:
The postman just sort of barfed up this big brown envelope outside my door, and while in the process of shoveling it into the trash, I noticed some of your feeble attempts at an address decorated on the outside, so, with an uncommon amount of courage, I opened it to see what might crawl out. Nothing crawled out except a few scraps of Light C-Grain Balsa (only a fish would wrap fiberglass around good wood. I guess the rest of the stuff was too busy being airsick after the flight in. They only bounced twice before getting the damn thing down). A Tech 7 is an update of the Tech V, using a new TI ic. It is pretty similar but doesn't use the external driver transistors of the V so is a bit cheaper to build for assembled servos and simpler in the case of kits. While the servo mechanics are not up to the KPS-14, 15 or the new D&R (RS-5) the amp is damn good. Super smooth. Definite improvements over. . . .

Your Royal Fishiness:
I have been doing some thinking about pitch control systems and flaps. As you may recall, we had a minor disagreement concerning flying

stab vs. fixed stab & elevator systems. In particular you wanted a flying stab for low trim drag and I wanted. . . .

Dear Fish:
Well, I am pissed off again. Here it is a good glider day (stationary Cu cloud over runway) and I'm sitting here writing a letter to you. Actually I am on watch again. Oh well, after long consideration (.2 nanoseconds), I have decided that you are probably afflicted with the rare disease Cantseeus Treeus. Unfortunately, I don't know of any way to build a glider that has the same disease. I don't like the idea of having a 40 percent flap. . . .

Fish:
You were bitching about me not sending you enough letters, so I figured I would send you another one. Enclosed is a schematic and pictorial of Cannon Tini Twin RX, including values for 53 MHz. The land layout is left as a trivial exercise for the student. Board is about 1½" square, to fit alongside of a mini servo mechanics. Note 8 channel decoder. I had a few problems with. . . .

Fish:
Received your latest randomness. I notice the institute has finally come to its senses and evicted you, probably so you won't lead so many poor defenseless freshmen and other random tools down the road to sin. I got the Cannon Tini Twin today. Servos are the new D&R with the same IC as the Royal Tech 7. Receiver is double tuned front end, with FET mixer, 74164 IC decoder. . . .

Fish:
I am sending your Christmas gift to Maryland. I suggest you open it outdoors and away from everybody else, as I wouldn't want to accidentally injure anyone. Thanks for the bolt. I needed one that size. Ye Olde Coastie Guarde has decreed that humble self shall go back to Pt. Arguello. Of all the Loran stations in the U.S. (of which I requested half) they decided to send me back there. In any case, when is spring vacation? If you are going to be home, I could . . .

The lengthy correspondence with Parky about boost gliders went into a box. They were just as important as any love letters he ever received, Gup said. They were just as valuable.

By 1978, Parky had finished his tour with the Coast Guard and he and Gup had finally perfected their boost-glider designs. After four years of detailed correspondence and three years together at the Institute, they had built radio equipment for models weighing less than ten grams, and rocket bodies that included a parasite glider with a twelve-inch span, also weighing less than ten grams. Gup's particular handiwork, called Dark Star, would be ready to go with them to Yambol, Bulgaria, in September that year, when they hoped the nature of international competition would take a serious turn toward the NATO alliance. The Communists, who ordinarily hired professional engineers to compete in space rocket matches, would be taken by surprise.

Back in Washington, John boxed up the new Athena H-003, the extraordinary scale model he handled with white gloves. Parky completed his scale model, made of Kevlar and graphite, which the judges complained was not a rocket but a jet. Although they entered the world championships with devices that perplexed the Warsaw Bloc, the team did very, very well. The MIT Rocket Society finished second in the world championships of space-modeling competition. And after ten years of trial and error, Guppy's performance with the Dark Star was flawless. They said when the Bulgarians presented him with the gold medal, he jumped off the winners' stand and kissed the native girls presenting flowers. John Langford snapped pictures to prove it.

4

They did not return to a heroes' welcome in Cambridge but to a list of course requirements and a tight housing market. The lease on Gup and Parky's apartment had run out while they were away, and faculty members clamped down on them to finish course work and graduate in the spring. Around the aero department, the rocketmen were largely ignored for their achievements and encouraged to move along through the academic system. The only good news was the National Air and Space Museum wanted John's scale space model and Parky's X-2 to put on permanent display at the Smithsonian. John gladly turned over his scale beauty, but Parky enjoyed flying models more than fame, and the X-2 completed its life cycle in swirls around Cambridge rather than on a shelf in Washington's rarefied air.

With the championships completed, no more excuses stood between the graduating class of '79 and the remaining rocketmen, John Langford, Bob Parks, and Hal "Guppy" Youngren. A slew of incompletes hung over Parky's head, particularly a set of humanities courses, which he despised not so much for their content as for the writing assignments. Parky loathed the process of writing. He could have aced music appreciation by turning in a ten-page paper, but he could not produce a single document. The same for archaeology, history, and English. His handwriting was abominable. His typing was splotched with errors. Inevitably he thought so much faster than he could write that often his sentences abandoned key words and phrases in a synoptic lurch toward a conclusion. "I found it so incredibly painful that it just never happened," he said. Also, since he had first entered MIT in 1969 course requirements had changed, so not only did he have credits from a few classes that didn't exist anymore, he had to make up for new core courses that hadn't existed when he was a freshman.

Gup's academic baggage was equally heavy. Although he'd entered the university with advanced credits in math and English, he'd toyed with an experimental studies program and sampled such a large variety of classes that it was unclear just what kind of degree he might have qualified for.

The world champs could no longer hide behind their slogan: "Bureaucracy über Alles—The impossible we do overnight, the paperwork takes forever." MIT's administrators bore down.

John, as usual, steamrolled through his remaining classes and continued to wonder what dismal career choices he'd face after graduation. Although he'd started college at least five years later than Parky and Gup, he wound up taking the standard aerodynamics class—unified engineering—with Parky, and 16.62, the aero department's one and only lab course, with Gup. In the mandatory airplane design course, Parks and John combined strengths, so Parky did the basic airplane design and analysis and John wrote the paper. In 16.62, he and Gup collaborated to build a device to gauge the thrust of model rocket engines. Gup built a data acquisition system—essentially, a sophisticated computer to convert the blast of an engine into pulses, calibrate them, and translate them into a series of numbers that defined the thrust. John's contribution was to build the test stand that held the rocket in place—again, a modest achievement compared with his mentors', but relatively easy to handle within a semester and check off the list of completed course requirements. He joined them in the dogged pursuit of a degree. It marked the end of an era.

One day in October, as John sat in the lab instructor's office putting strain gauges on his rocket test stand and wondering what to do with his life, a graduate student named Ed Crawley came in to make a phone call. The aero department's only full-fledged airplane project sat out at Hanscom Air Force Base, a withering hangar queen. It was a human-powered airplane called BURD (Biplane Ultralight Research Device), and even though faculty members and students had mulled over it for ten years, they'd never gotten it off the ground. The management at Hanscom wanted the airplane out of its hangar, and Crawley, who'd spent years building and studying the machine, was now trying to encourage a museum in Maine to take it off his hands. After listening to the phone conversation, John concluded that the museum didn't want it and the BURD would be dismantled.

John thought the notion of human-powered flight was particularly insipid and that the BURD had been a time sink, an aeronautical nightmare designed by a bunch of guys who didn't know how to build, dreamed up by people who knew theory but couldn't find their way around a C-clamp. After Crawley hung up, John made his pitch.

"Hey, Ed, if you're just going to tear it up, why don't you give it to the rocket society and let us put some model airplane engines on it, because, well . . . I bet we could fly it."

Ed, who mostly needed to return to his doctoral work, was delighted.

At dinner that night over in Walker Memorial, John presented the idea to Gup and Parky. Gup, who had once applied to be the BURD's test pilot and was turned down, thought it would be an interesting hack and, if it worked, might give him a chance to fly for the first time under his own power. Parky, who really liked the idea of human-powered flight, saw a chance to play with a new concept and savored the thought of redesigning the BURD so she'd lift up under the rocketmen's power right out of the inexperienced hands of MIT's theoreticians. John called Crawley and arranged a visit with the queen.

When John and Gup and Parks drove out to the dingy hangar a few days later, they found the device in an abandoned corner of the dirt-encrusted floor. It was the most bizarre real airplane they'd ever seen. BURD had an enormous wing span of sixty-two feet and stood fifteen feet high. It weighed only a little more than one hundred pounds. Its body was tightly wrapped in transparent polypropylene sheets so they could actually see the intricate skeletal structure made of balsawood trusses, blue Styrofoam particles, braided steel cables, Kevlar fibers, and aluminum tubing. A fancy two-seater Italian bicycle extended from the center of the long, crystalline, bi-level wings, and attached by a tube, six feet from the front wheel of the bike, was a single mini-wing for flying stability—a canard! The airplane was an old-fashioned, stabilizer-first canard. But not only that, not only had the BURD's designers outfitted the machine with its stabilizer in front of the wings rather than at the rear of the fuselage, they'd also put the propeller, a fifteen-foot-long laminated block of balsawood, behind the wings. A pusher prop!

They touched the flat tires, tweaked the broken steering mechanism, and tugged at slack control cables. As they ran their fingers along the airfoil, they bumped up against one broken rib after another. The BURD looked like a tattered scale model blown up beyond its true size, exaggerating its defects and imperfections. They sort of liked it, actually. The BURD was a discovery in hyperbolic sizes and weights and materials and new construction methods. Best of all, they liked to think they could make it fly. After ten years, this monstrosity was the best flying machine the department of aeronautics could construct. And such a kludge as they had never seen.

In fairness to Professors Eugene Covert and Jim Marr, the BURD team's faculty advisers, few people in the world had ever been able to build a successful human-powered airplane. The first human-powered flight in America occurred thirty-five hundred years after Daedalus made the first storied flight. In 1976, a retired Air Force colonel from Rhode Island pedaled an airplane—a 291-pound monoplane that took four and a half years to build—seventy-seven feet through the air. It flew for five seconds, and it only flew once. Throughout the twentieth century engineers in Germany, Great Britain, Japan, and France had applied their knowledge of aerodynamics to build airplanes that could be sustained by human power alone. They hired professional cyclists, experimented with designs that ranged from flapping batlike wings to gliders, engaged professors, sponsored prizes, and even courted the interest of their governmental leaders. (Germany's Air Marshal Hermann Göring once personally donated 3,000 marks to the Frankfurt Prize for human-powered flight experiments during the 1930s). Still, they measured their "flights" in hops, skips, and jumps. The best engineers could not divine materials light enough and strong enough for a single prolonged flight. The best athletes could not provide enough power. After much trial and error, distinguished scientists left the world a legend of goofy film clips which captured their awkward inventions humping along runways, being heaved off cliffs, and always, inevitably, disintegrating under duress.

At the time of BURD, the world's sole existing prize for human flight was offered by the Royal Aeronautical Society (RAeS) in Great Britain. An English industrialist, Henry Kremer, had pledged $129,000—the largest prize in aviation history—to the first person who could fly a heavier-than-air machine, solely by human power, around a figure-eight course embracing turning points not less than half a mile apart. The RAeS made the offer first in 1959, and for eighteen years no one could claim it. In August 1977, a very professorial entrepreneur, Dr. Paul MacCready, Jr., and a team of professional engineers and aeronautical hackers from California, flew the Gossamer Condor 1.35 miles around a figure-eight course. Some of the BURD's builders thought the Condor looked an awful lot like their airplane, and when MacCready was credited with making a revolutionary leap in human-powered flight, they kept quiet but privately scratched their heads. "MacCready liked to tell this story about how he was driving across the country and had done some power calculations on hang gliders," Crawley said, "and if you made this low-speed, wire-braced, canard airplane, it could fly. I'm sure there's some truth to it, and he's a friend of mine and he's a nice guy and all that, but he was also very knowledgeable about the BURD project, which is a low-speed, wire-braced, canard airplane."

Nonetheless, the BURD did not win any awards. It did not attract a great deal of attention. It did not fly. By the time the rocketmen reached it, Professors Covert and Marr, the men who'd overseen the BURD's birth and rapid decline, spoke of it as an educational tool, and the lab instructor, Al Shaw, used it in his Tuesday safety lectures as an example of things that can go wrong during flight operations. The airplane was good for nothing more than an anecdote or two.

During the final flight attempt of the original BURD, Al would tell his students, a camera crew joined the team at dusk in a station wagon to film the flight, and when the airplane's big white wings flipped up, tip to tip, during its lunge for takeoff, the driver of the station wagon mashed the brakes. One team member, who was on the roof of the wagon, and the entire camera crew, tumbled helter-skelter; the guy on the hood hit the pavement and nearly slipped under the wheels. Worst of all, the cameraman didn't capture the moment when, for a few microseconds, the BURD reputedly lifted off the ground.

Parky did most of the calculations sitting on the hangar floor. After running through the numbers on power and stability requirements, he concluded that the thing was, essentially, a disaster. The control authority needed to be quadrupled; the original idea of using two people to pedal seemed a waste since all of Parky's numbers suggested one pilot would be more efficient. In conference, Parky and Gup decided to eliminate human beings as the engine and install real power. They mounted a couple of two-by-fours on either side of the fuselage, and bolted on a set of red model airplane engines with twelve-inch props good for 1.5 horsepower each at 15,000 rpms. For ease of operation, Parky rigged up a radio-control system for the engines. To empower rudder control, they designed a system of separate throttles on the two airplane engines that gave them enough authority for turns—just power up one engine and let the other idle.

They taxied in mid-October. At first they taxied with engines only. The little devils screamed and spit out seven to eight pounds of thrust each, but the BURD would not fly. The BURD wouldn't even roll along the ground at more than seven miles per hour. Then they put Gup in the front seat to pedal the big prop and provide another half a horsepower. The BURD rolled faster down the runway but it would not fly. The rocketmen did more calculations and decided the main landing gear had been built in the wrong place and the airplane wouldn't rotate—that instant when the nose first lifts up—because it was nose-heavy. Gup moved to the back seat and they rearranged the engines and controls, and although they expected the configuration would make the BURD unstable, Parky told Gup it was his problem to learn how to fly it once he got in the air. And still the BURD would not fly.

Finally, just before Thanksgiving, they decided that if they could get a long bungy cord and hook it up to the back of a truck, put Gup in the pilot's seat, and run the engines at full blast, then maybe the plane could pick up enough speed to lift off the ground. With thirty pounds of thrust humming down the runway at twenty-nine feet per second, they figured, even a stuffed goose would rotate.

A couple of dozen people showed up at Hanscom one cold November morning to watch. Under a gray, marbled sky, they took the BURD off its cinderblocks and rolled it out of the hangar, past a line of helicopters, past patches of snow, down to the white landing strips at the edge of the runway. In the glare of the rising red sun, everyone crowded around the BURD.

Al Shaw held a video camera. Ed Crawley climbed into the truck, and a couple of professors looked on. Hyong Bang snapped pictures. The rocketmen wore their USA hats and jackets from the world championships. Everyone's girlfriends were there. Parky tested the engines with his remote control and Gup climbed into the back seat wearing a set of ear protectors so the whining wouldn't crack his eardrums. Parky punched his controls and Gup started to pedal.

The truck lurched forward. Gup pedaled harder. The engines screamed. A crowd of people burst down the runway after the hurtling BURD, and just as the airplane reached its theoretical flight speed, Ed looked back from the truck and he saw the nose lift. There was a gasp from the canard, a crack, and then the BURD began to skid. And teeter. It crumpled up and fell over in a heap.

Everyone stood around for a long time, looking at the wreckage. Parky conceded, "The airplane may have flown." Technically, he said, if the springs on the main wheel were relaxed when the BURD rotated, maybe you could call that flight, even though the wheels never left the ground. Ed said he felt a personal sense of relief because the project had ended cleanly and now he could return to his doctoral work. But the rocketmen sensed that he was genuinely depressed. They swept up the litter and rolled the BURD's remains back to the hangar, down past the yellow, frozen fields and the helicopters. Then they stopped and everyone gathered into a group and took off their mittens and caps to have their picture taken in the shadow of the dying BURD.

"You know," Ed said, speaking to no one in particular, "it's really not so bad."

"What do you mean, it's not so bad?" Parky said.

"Well, in a thousand man-hours, you could build a new canard."

One thousand man-hours! Gup looked at Parky and Parky looked at Gup. He couldn't be serious. A thousand man-hours? That was an insult. In a thousand man-hours, they could make four new canards. They could construct a decent set of wings. They could design a real flying machine.

"Ed," Parky said. "In a thousand man-hours, we could build a whole new airplane."

The Friday after Thanksgiving, the Model Rocket Society members went back to the hangar, and with hacksaws and axes and hammers and steel

Rafter squares, they assumed responsibility for disposing of the BURD. They piled the remains in a heap and Parky jumped up and down on them like a kid in a pile of leaves. When they were finished, they stuffed the broken bits into trashcans and pledged to build their own human-powered airplane by springtime.

5

In December 1978, Bob Parks found himself spending a lot of time on the third floor of Walker Memorial. Running continuously between the rocket society offices, where he crunched numbers, and the aero department library, where he searched for technical papers, Parky collected performance data from as long ago as World War I and discovered what made the BURD not want to fly. In the hunt for old ideas he had epiphanies for human-powered flight. He thumbed through "The Prospects for Man-Powered Flight—The Future of an Illusion," by John McMasters and Curtis Cole; "Airfoil Design for Man-Powered Aircraft" by F. X. Wortmann, University of Stuttgart; "Human Power Production in Relation to Man-Powered Flight" by Professor D. R. Wilkie, Department of Physiology, University College London; and "General Biplane Theory" by Max Munk. The prospects for building an HPA, as they'd started calling their human-powered airplane, seemed increasingly good. Before his friends left Cambridge for the holidays, they'd talked excitedly about building a plane to compete in the new Kremer competition, a $200,000 prize for the first HPA to cross the English Channel, and they christened the new venture "Cross-Channel Enterprises." John and his fiancée, Barbara Jenkins, designed T-shirts with that logo.

The problem with BURD, Parky decided, was it had been built by scientists, not engineers, and suffered the common scientific affliction, Paralysis Through Analysis. Despite the many academic articles written about the BURD, the wind-tunnel tests and mathematical pursuits, fundamental deficiencies showed in its engineering. Not only was the airplane's center of gravity too far forward, but the spar had broken under stresses about one ninth of what it was designed to support. In the early designs, the wire trusses bracing the bi-level wings were quite strong in the plane of the wings, meaning that the designers had thought seriously about the forces the airplane would meet in terms of drag. But they'd hardly considered the forward tug of lift loads (the pressures of lift) that would assault the wings—the wire bracing in the direction of the wings had been completely inadequate, and thus, during an early flight operation, the wings flipped forward and touched tips. "It's real simple," Parky said. "It's real stupid."

What he proposed—in concert with Gup, of course—was to employ fundamentals of airplane design, elementary model airplane lessons. Rather than spending an inordinate amount of time examining new airfoils and creating hypothetical designs that would take forever to build, they'd construct a humble, very conservative and practical model airplane first, as a prototype. After years of rocket building, they'd learned to stay on this side of too far. They needed something about one-eighth scale. Something they could test for controllability. Something that would fly from the start. Different faculty members advised them to conduct wind-tunnel tests and computational airflow analyses. "Sure," Parky would say, "and then maybe two years later we'll come up with an answer. Maybe. The other approach is to go build it and try it."

The plane they'd call Chrysalis would not be a flying wing, it would not have a pusher prop, it would not entail a canard. The tail would go in back, the propeller would go in front. There would be one pilot. The only issue was whether it would have one wing or two, and that matter was easily settled by the fact that the hangar at Hanscom was only seventy-five feet wide. Chrysalis would be a biplane with seventy-two-foot wings. They would build a scale model, test it, and if it flew, they would build it to size and fly that one, too. Within days Parky would design a very ordinary human-powered airplane by relying on the basics. After years of modeling experience, Bob's Basics were these:

1. Part of the art of being a good engineer is knowing where to steal good ideas.
2. There is no perfectly optimized design because you cannot account for every aerodynamic effect on an airplane. Get it good enough, Parky would say, and get on with it.

3. The best way to solve an engineering problem is to guess the right answer.
4. Any damn fool can figure out another way to do it; i.e., the first solution to a problem is often entirely adequate.
5. If you can't fix it, feature it. (Parky found this one on a calendar once. If a design problem can't be turned into an engineering advantage, there's always a way to market the problem as a benefit.)

Proper-thinking faculty members choked on Bob's Basics. The trend in the industry since World War II had been away from workshop engineering, designing airplanes from scratch, building prototypes and having them fly within a matter of months. Instead, the task for an engineer—an MIT engineer—was to foresee every possible contingency. Aeronautics had shifted from an engineer's habitat into a strictly defined scientific arena, yielding an emphasis on developing new theories and equations to describe such matters as fluid flow, which couldn't be solved until someone invented the math to do it. At MIT the department had become so atomized—you were a propulsion expert or a structures expert or an airfoil expert or a computational flow expert—that no one really bothered to integrate the systems or toy with the simple matter of producing an airplane. That's why Parky called MIT "a prep school for Ph.D.'s." It's also why so many of the rocket society members felt ambivalent about their futures, caught between the theoretical trappings of academe and the inertia of the airplane industry. John Langford, for example, knew full well that the last airplane to go from drawing board to runway at Lockheed in Atlanta was the C-5 Galaxy, a project started in 1963. Screw all that.

One afternoon, rising from his chair to get something to eat after days of unrelenting work, Parky suddenly realized it would be damn hard to find a restaurant open. It was Christmas Day. True to form, the team was at work. Parky was mulling over the final design details. Gup was at home in Maryland deciding on the structural layout. John was in Atlanta, working out a budget proposal to the department chairman. Hyong Bang had begun working on an optimized propeller design. In the weeks that followed they would shell out fifty dollars for a set of plans for Paul MacCready's Condor, and steal into the National Air and Space Museum carrying binoculars to look at the Condor for construction ideas. They'd build a one-eighth scale model of the Chrysalis and fly it around the DuPont Gym. They might not make breakthroughs in science, but they'd sure teach the department a lesson in basic engineering.

"In terms of aerodynamics," Parky figured, "we don't need any new knowledge. You just put a big wing on it so it will go really slow. Get enough structure to hold it together, enough control surfaces to control it and possibly make it stable. Put a big prop on it, and pedal like hell."

Parky made everything sound so easy. In a casual, slightly high-pitched voice, he could unshell the nastiest aerodynamics imbroglio, pausing only to comment that "the problem is non-trivial," or, "this is fairly amusing," as if to remind himself that engineering is a game, after all. Whether juggling a set of performance equations or shaving a few ounces off the fuselage of a balsa model, Parky's approach to airplanes was always the same—be practical.

Like John and Gup, he seemed to have been born with a natural curiosity about flight. At approximately age three and a half, he glued together his first model airplane while his parents were on vacation, and it became an obsession almost immediately. Somehow, he'd retained that attitude toward airplanes as a plaything, as if every aeronautical challenge was just another neat box of plastic parts.

As a child growing up in Lakeland, Florida, he could stand in his backyard and see the flashing fireballs from rocket launches at Cape Canaveral, one hundred and twenty miles away. When his mother went to bowling tournaments around Cocoa Beach, she'd take him to visit the space museums. But the motivation didn't come from NASA. Nor was it inherited from his father, a salesman who worked up to one hundred hours a week as a broker in the citrus business and who set aside his spare time for outdoor sports rather than workshop hobbies. And despite what Freudians might surmise, his immersion didn't stem from some innate desire to fly. Parky never particularly enjoyed flying. He liked building models. "I don't know where it came from," he'd say. "It's just that way."

By the time he was six years old and had filled up his bedroom with models and stocked the laundryroom with models and bought his first model airplane hobbyist magazine to study, his mother and father began to worry.

Every horizontal surface in their home was covered with his handiwork. When he reached the third grade, they decided to negotiate.

"Bob," his mother, Mae, would say, gently, "your father and I think you need to spend more time on your homework. No models for six weeks." Bob's grades would drop for six weeks, and when his parents relaxed the rule, his grades improved. His mother tried another tack.

"Bob," she'd say, "these airplanes have cluttered up your room. From now on, if you want to buy a new model, you have to get rid of two." For a while, Bob would relinquish two models so he could build another one, but his output increased so prolifically that the house stayed cluttered with Dremel tools and glues and scrapwood and paints and, again, plastic models. He just seemed to build more and more.

By age eight, he'd also tracked down Lakeland's model airplane club, the Lakeland Balsa Termites. The adults who ran the group saw his talent and passed a special resolution on age requirements to let him join. He also made friends with a retired TWA pilot and former barnstormer who lived down the street, and they would spend hours talking about airplanes. His parents finally learned to tolerate Bob's "interest." He was an extremely good-natured child, after all. He made excellent grades and scored near-perfect marks on achievement tests. He wasted no time watching television. Other than this one peculiarity, he was perfectly normal. Rather than try to stop production, his mother discovered a way to encourage the hobby and yet slow the proliferation.

"I set him up with a charge account at a store here in Lakeland called Toy Town," Mrs. Parks said. "So he'd take the bus from school to the store and then walk home. He was allowed to spend any portion of his allowance, but if he found something he wanted that was more expensive, he could charge up to fifty cents. Some friends of ours owned the store, so they made this arrangement possible: the deal was he couldn't charge again until he paid off that fifty cents. And it worked. He'd spend hours walking up and down the rows looking at models with his hands behind his back."

The lessons he learned from little plastic airplanes, and later rockets, embraced a variety of sciences. By the time Parky entered high school, he knew more about physics than the high school's physics teacher, a man whose primary job was to coach football, and the principal asked Parky to teach the course. When the public address system at school failed, they asked Parky to fix it. He rigged up remote-control devices in a storage room at home to make an all-electronic photographic darkroom. He won a science foundation scholarship one year to attend the Illinois Institute of Technology, and after a summer's work, he knew far more about computers and computer programming than anyone in his school. He graduated at the top of his class, and won enough scholarship money to fund much of his education at MIT.

He grew tall, thin, and slightly stoop-shouldered; in very palpable ways, concentrated work shaped him. The attention to minute details, the quest to fly things lighter and smaller and faster, bowed his back. By 1978, Parky's talents were obviously destined to be used at a drawing table and workbench. Despite the energy he gave to models, Parky's craftsmanship was still not quite as extraordinary as Guppy's and his ambitions were never nearly as bold as John's. But his mind leaped ahead of nearly everyone's and, perhaps just by dint of personality, he had a knack for harmonizing imperfections with workable solutions. Guppy and John agreed that Bob Parks understood engineering intuitively. When the time came to design MIT's first workable human-powered flying machine, Parky approached it like just another model airplane.

A human-powered airplane operates in a peculiar realm, a flight regime more comparable to a goose or a pigeon than a Cessna. Because its energy comes from a biological source, whose power is finite and relatively measly, these airplanes fly low and slow. With maximum human power set between a quarter and a half horsepower, the pilot must necessarily rely on long and efficient wings designed to manipulate the airflow around them for a generous amount of lift and very little drag. The wings must not be so heavy that the person, above whose head they balance, finds them burdensome and finally wastes more energy straining to lift them than lifting his own weight. The whole puzzle of human-powered flight, in fact, revolves around this balance between low weights and low power, and as engineers begin struggling with this quandary, they continually bump up against the same limitations. In the end, the trick becomes making a craft that needs to be about half the weight of its pilot, strong enough to withstand all kinds of pressure, loads, and torsional effects from the air pressure, wind, and the body inside it, and that will float through the air—just on the verge of a stall—in perfect trim or balance.

In the past, most engineers had approached the problem by trying to incorporate their designs with principles learned from airplanes with the lowest drag—sailplanes, essentially, motorless gliders with propellers. The results were nearly always the same: eccentric little streamlined, low-drag, single-wing airplanes demanding high flight speeds and so much pilot exertion at the pedals that little attention could be paid to the controls. If they flew at all, such airplanes flew hysterically and rarely survived their crashes.

Examining separately the matter of wings, structures, stability, propellers, and power, the Chrysalis team made practical choices that any aero modeler could understand. For example, by using biplane wings, they not only increased their effective wingspan (and lift) beyond the seventy-five-foot width of the hangar space, but they also gained a much stronger structure with the biplane's crisscrossing trusswork between the wings. Their biplane with seventy-two-foot wings would equal the performance of a single wing at one hundred three feet, and be as sturdy as the Harvard Bridge. As odd as it seems, two wings would also be lighter than one. A biplane's external wires and struts could absorb the stresses of flight much better than a single, heavy, sturdy spar or series of spars in a monoplane. What they lost in terms of dragginess and aerodynamic perfection, they gained in significant weight savings.

The airfoil, the crucial surface design of the wing that keeps an aircraft afloat, presented another problem they approached with an attitude of shameless pragmatism. Usually, an aerodynamicist would look for a cross-sectional shape (i.e., airfoil) with a high lift-to-drag ratio, a streamlined contour the shape of a teardrop, subtly crafted to distribute the air pressure over the wing so the flow would be exceedingly smooth. Ideally, the layers of air skimming across the wing should be laminar, that is, coursing over the foil with as little turbulence as possible, since turbulence causes friction between the air and wing surface, and friction produces drag. The airfoil should also have a low pitching moment, meaning it needs to be shaped in such a way that it will remain level along the axis of flight. As the name suggests, a pitching moment is a force created on a wing that tends to twist the wing forward, and unless the wing is strong enough, a large pitching moment can redirect an aircraft with as much force as any major control surface, an elevator or rudder. Handling a serious torque around the axis of the wing requires larger tail surfaces to offset the pitch or stiffer materials to hold the wing in place. Either way, correcting a large pitching moment adds weight to a flyable airplane. Above all else, in a human-powered airplane, more weight does not fit into the equation.

Inspired by Bob's Basics, the Chrysalis designers did not attempt to develop a superior, low-drag, high-lift airfoil with a small pitching moment. They stole the one from Paul MacCready's Gossamer Condor. The Lissaman 7769 airfoil, with its arched top and gently pouting bottom surface, was not the perfectly sublime airfoil one might expect in a finely tuned HPA. Quite the contrary, it was highly turbulent and excessively draggy. But it was also much less risky than an all-laminar airfoil. The Lissaman 7769 could tolerate sloppy construction, it would be unaffected by surface grunge, and it could be built cheaply and quickly. It wouldn't fly as fast, but it would fly, and it did have the characteristics of high lift and a low-pitching moment. A more perfect airfoil would require tapping deeply into the physics of laminar flow theory, and even then, such an airfoil would be finicky. In accordance with Bob's Basics, since they couldn't fix it, they featured it. The Lissaman would fly just fine.

For stability and control, they sought counsel with the Wright Brothers. Instead of adding ailerons—hinged control surfaces at the rear of the wing used to bank an airplane—they avoided the extra weight by employing the ancient art of wing warping. Essentially, by putting a few pulleys in the right places and leading a stretch of piano wire through some guides from the cockpit out along the top wing, down the struts to the bottom wing, and back to the fuselage, the engineers devised a way their pilot could wiggle the wings by pulling a lever. The wings could turn side to side for a different angle of attack at the air, and the airplane would be able to bank for turns. Many of the pioneers of flight used wing warping, including the Wrights. All that was needed was a way to stabilize the airplane when it banked. By designing the bottom wing so it angled up slightly from the center to the tips, the engineers easily employed another old-fashioned idea to create lift on the wingtips when the airplane rolled. This so-called dihedral provided a strong corrective moment against sideslipping, that moment of instability when a turning airplane could lose its balance in the air and take a quick sidelong slide. Again, the solution was inherent to the basic design and did not heap additional weight on the pilot.

Hyong and Guppy worked on a uniquely designed propeller that was the creation of their faculty adviser, Professor Eugene Larrabee. The rocketmen had asked Larrabee to be their official adviser primarily because he was one of the few remaining persons in the department who still seemed to care about building airplanes and yet remained busy enough with his own research not to bother them. He also taught biplane theory and knew a lot about propellers.

A propeller can be just an outsized fan with twisted planks for blades, a crude propulsive force, and still work. But carefully considered, it's apparent that the blades operate under the same basic principles as any other airfoil. By focusing on how to distribute thrust (or lift) over as much of the span as possible, an efficient propeller can be devised that will serve as the most important aerodynamic surface of an airplane. What Larrabee proffered to the Chrysalis team was the original theory of minimum-induced-loss propellers developed by Albert Betz and Ludwig Prandtl, two German aviators, in 1919. The theory had been largely ignored since World War II until Larrabee revived it. He transformed it into a convenient form for computation, and Hyong and Gup coded up algorithms for the actual design. They wanted to make the propeller as large as possible and yet still have some ground clearance. The choice was to have fast-turning, long, skinny blades or relatively slow-turning, fat blades. Those decisions rested on the kinds of gearing ratios that would fit inside the fuselage. As usual, weight limits defined their choices.

With all the parameters set for gearing and ground clearance, the number of blades, and power requirements from the pilot and the prop, Gup and Hyong loaded the computational formulas onto a computer program, plugged in the numbers, and out popped the acceptable shape and twist of the Chrysalis propeller. The most sophisticated part of the airplane had grown out of a forgotten theory devised at the end of World War I. Designed to be fourteen feet long, made of Styrofoam, supported by an aluminum root spar, and covered with Kevlar and epoxy, the prop promised to work with close to 90 percent efficiency. Its projected weight was just under three pounds.

By early January 1979, Gup, Hyong, Professor Larrabee, and Al Shaw joined Parky at DuPont Gym on the MIT campus to watch him float a scaled model of Chrysalis around the basketball court. While everyone gawked, Parky took a few strides across the court, loping like a giraffe on his long, skinny legs, and gently tossed a lustrous aircraft—balanced with nine-foot-long, transparent, bi-level wings—into the air. Parky twisted a set of knobs on a hand-held radio-control device and tested wing-warp, rudder, and elevator controls on the one-pound plane. The motor, adapted from an SX-70 Polaroid camera, purred. They watched the model wiggle and bank between half-court and the free-throw lines. Someone commented about the slow roll response in turns and Parky calculated how much more dihedral the real airplane would need. Pick-up basketball games stopped and sweaty players turned their attention to the rocketmen, who kept launching this weird craft and crashing it, launching it again and watching it shimmy down court.

After several days of eyeballing their experimental model at DuPont, Gup and Parky concentrated their efforts on an improved fairing—a more streamlined form—for the fuselage. They built a scaled version of Larrabee's prop to replace the standard model airplane propeller, and decided to rely on a combination of dihedral and rudder control for making turns rather than using wing warp. They made corrections on the model. Then, one night, when the gym was empty, Parky and Gup went to DuPont and flew the plane around the court three times in perfect figure eights, just like a miniature version of the first Kremer course. They flew it until the batteries in the Chrysalis model gave up their last volts and the airplane finally floated to the floor, undamaged.

On February 15, John Langford visited the chairman of the department, Professor Jack Kerrebrock, and asked him for $2,500 to build a human-powered airplane. John showed him a budget, a schedule, some performance numbers, and a list of goals:

1. The airplane will be operating by May 1979.
2. The airplane will be built so any member of the flight crew or faculty can fly it.
3. The airplane will be semi-portable and easily stored in the available space at Hanscom.
4. The airplane has been designed for easy construction and repair.

5. The airplane will provide growth options/technology readiness, as the first in a possible series of crafts which may attempt the crossing of the English Channel, as set out by the rules of the latest Kremer Prize.

Kerrebrock looked at John's expertly organized plans, the ninety-day schedule, and the budget proposal. Then he looked at John, who was among his best students and who had just returned to MIT after spending yet another month at the White House Office of Science and Technology Policy. Was this just another crazy kludge from the rocket society, or was this young man serious? The chairman wouldn't truck with foolishness. He had no time for ineptitude.

"How do I know it will fly?" he asked.

John reached into his pocket and slipped out a Polaroid picture of Bob Parks flying the one-eighth scale model. He set it down on the desk. Within hours, the radical rocketmen were once again in business.

6

Unlike most classrooms in Building 33, the aeronautical engineering lab existed entirely at ground level. Set essentially in the basement of the aero department, the big room had all the attractions of a well-heeled hobbyist's workshop. The musky scents of sawdust, oil, ground aluminum, and melted Styrofoam flavored the air, and a shrieking chorus of drilling machines and sanders, jigsaws and routers, straddled the rough wooden floors. Don Wiener ran his sector of aero-astro with the discipline of a shop foreman and the seriousness of a small businessman. His lab didn't stand on par with the reputation of MIT's theoreticians, but it was still the cradle of engineering, the only place where undergraduates confronted the blade.

Don had seen plenty of smart kids come through the shop thinking that if their projects needed a sheet of aluminum with a hole in two corners and one in the middle, all they had to do was find the right specialty store in Boston. "No, son," Don would say patiently. "You're gonna learn to drill those holes yourself." And before the semester ended, he'd have them off in a corner somewhere making the best-looking objets d'engineering that could be coerced from them. Most students floated through, fulfilled their simple requirements, and left behind competent but uninspired engineering products. Consequently, Don maintained a sense of order over the place and didn't have to worry about the property being misused, overused, or even used after hours.

One Monday morning, though, not long after the Chrysalis team made its formal presentation to the faculty, Don came to work and noticed something awry. The milling machines were locked behind their wire cages, just as he'd left them on Friday. He looked in his office. There was the SKIL calendar with the picture of the buxom girl in blue-jeaned overalls, the usual stacks of parts catalogues, the leather-bound volume of *The History of World War II*, his stained coffee cup with the Empire State Building decal—all undisturbed. Maybe this place just looks too clean, Don thought. At the end of the week, he secured the locks before he went home. When he returned on Monday, the shop still didn't look right. Too clean? he thought. No, maybe it's too dirty.

After a few weeks, Don decided he had a problem with intruders. He rigged up a device on the doors to the milling machines so even if someone could unlock the cages, the latch wouldn't lift. He thought he'd outsmarted the gremlins. On Monday morning when he came back to work, everything looked fine again, except for a few tool bits. They seemed different somehow. He stared at the titanium and carbide tips, shiny with the glint of a new taper. They'd been ground into odd configurations, with racey slopes and angles. Don thought, Jeez, what the hell happened to this tool bit?

And so it was that the Chrysalis airplane took shape.

All it took was a little hackers' cunning, and they were in. With Don's shop clandestinely available to the team on weekends, staggered episodes of construction became possible. Every Friday night they rearranged all the machines to make construction time more efficient, and moved all the student projects under the tables to keep them clean and out of harm's way. By early Monday morning, Don would find his shop quiet, clean, and usually without any evidence of disruption. A few times, his students' 16.62 projects were inexplicably and remarkably improved at the beginning of the week—despite their deadlines, a few members of the Chrysalis team couldn't bear the sight of sloppy work. Don never identified the ghosts.

The self-imposed constraint of building an airplane in ninety days grew out of the boast to Ed Crawley that they'd make a flyable HPA in a thousand man-hours. While John went on a campaign to recruit, entice, and persuade undergraduates to help them build the wings, Parky and Gup set out on a difficult search for the right materials and jigs with which to make them. They'd already set parameters for new ideas in terms of the Chrysalis design, and now they faced the discipline of deadlines and available materials.

Gup's structural design depended on fifty-four thin-walled, high-strength aluminum tubes, which would form the lightweight skeletal frame of Chrysalis. He and Parky phoned all over the United States looking for aluminum tubes fifteen millimeters thick, but no one made parts that flimsy. Late one Friday afternoon they discovered a production company in Los Angeles with twenty-two-mil tubes. They jotted down the lengths and prices, and asked the manufacturer to put a freeze on his supply. Over the weekend they gauged the company's figures against their design and discovered that their single order would wipe out half the Chrysalis budget and deplete 35 percent of the country's entire stock of twenty-two-mil aluminum tubes. Gup and Parky weighed the variables—and early Monday morning they called L.A. to make the purchase.

The major material throughout the airplane was plastic foam, used for wing ribs and leading edges. Known to housing contractors as an insulating material called beadboard, the long blocks (eight feet by two feet by two feet) posed a problem because someone had to cut dozens of sheets as thin as one sixteenth of an inch—by hand. Two people using a hot wire stretched between guides could slice one-quarter-inch sections, but the process was extraordinarily slow. John proposed spending some money to throw "pizza parties" in Building 33 every Saturday, as a way to attract volunteers to cut beadboard. Every Saturday a dozen or so students would turn out, but since so few of them knew anything about building, the Chrysalis team members spent most of their time explaining how to handle various tools. Usually the parts they made were seriously flawed—from ripply, slightly gouged textures on the leading-edge sheeting to ribs that didn't quite match their templates. For novices, modest imperfections might have been passable. But even on the forgiving Lissaman airfoil, a "modest imperfection" was a relative value; they couldn't allow technique to be quite so compromised. If the Chrysalis team insisted on good workmanship, someone had to find a better way of building pretty soon.

During the third or fourth pizza party, while John patiently explained to a new set of volunteers the importance of precision, Parky disappeared. Watching their supply of beadboard fritter away into useless cuttings, he grew impatient and soon began rummaging around the lab for a workable solution. Within a couple of hours, he rejoined the group with a machine made of rollers and an electrical wire, a Teflon-coated ramp, and a two-pound weight tied to a line laced through a couple of pulleys. He attached the parts on a shop table and clamped a hunk of foam onto the ramp, set the gauge for a thin cut, and switched on the hot wire. When he released the weight, the foam inched slowly through the gate. Ten minutes later, the weight smacked the floor and everyone examined the slice. "Goowaaah!" John said. The first automatic HPA foam cutter joined the production line.

Most of the foam was cut and ready for assembly when the aluminum tubing arrived from the West Coast on March 13. The process of etching tubes into thinner sections didn't require many extra hands, but the Chrysalis team needed at least one more person to help who wouldn't be squeamish around potentially volatile materials. A quiet young freshman who'd shown up at one of the first pizza parties, Mark Drela, volunteered for duty. The team dressed him up in a gas mask, face shield, gloves, and a rubber apron, and prepared him for initiation.

On the third Saturday in March, a group of them gathered at 9:00 a.m. in the parking lot behind Building 33 and set up an outdoor chemical-milling operation. Parky taped up a sign saying: "Strictly Enforced No Smoking Area," and Mark started mixing chemicals. They used a stainless-steel soup kettle from the department's surplus bin to stir up a lye solution, and four hotplates to heat the water. But four hotplates would not heat water to 170 degrees. Everyone stood around in their gas masks wondering what to do. Parky snuck into the lab, and in a short while John saw him running into Building 17A with a Y-connector. Parky had discovered the steam-heating lines. Within a few hours, Mark stirred up a simmering tub of hot lye. Parky also thought to line the shipping cartons that contained the tubes with Styrofoam and plastic to make two long tanks into which to dip the poles. They dunked a tube into Mark's hot lye bath to etch away the aluminum, washed it off with a garden hose, and then dipped it into a solution of nitric acid, sulfuric acid, and sodium dichromate to strip off the alloys, copper, and silicon. The release of hydrogen between dippings could have caused an explosion. But by 9:00 p.m., when Mark cleaned acid off the last of fifty-four tubes, the worst damage appeared to have been done to Parky's clothes. A few days later, Gup measured the tubes with a micrometer and figured the one-day operation had already made their airplane eight pounds lighter.

John went ballistic. The project generated so much excitement that he quit fretting about his career dilemmas and chided himself privately in his journal for a tendency "to slip into thinking too much about the future when perhaps I should be celebrating the present." Just before spring break, John, like Gup and Parky, dropped a few courses—art at Wellesley and advanced mathematics at MIT—to spend time with Chrysalis, and he set out to search the campus for a large space in which the airplane could be pieced together. Only two months stood between him and graduation, and he had no notion what career path would open. All he knew was that nothing could be as exciting as building an airplane in ninety days. He'd set strict deadlines for construction and the team hit his pace.

They compared themselves to Kelly Johnson's legendary crew at Lockheed's "Skunkworks" in California, at one time the nation's fast track for aeronautical engineers. It was said that at Skunkworks, 23 engineers built the U-2 in 80 days, 23 built the XF-80 in 143 days, and 41 built the JetStar in 241 days. The story was that while Kelly Johnson ran the business, any airplane under construction wasn't allowed to be more than one hundred fifty feet away from the engineering team's desks. Unlike the rest of the industry, which customarily put a thousand engineers on contract to design a single airplane and spent years dawdling over construction, the Chrysalis team, like Kelly Johnson's, would prove that the impossible could happen overnight. The combination of John's management skills with Parky and Gup's technical intuition sped the Chrysalis team toward deadline.

They were all reading a book about Skunkworks when John struck a deal with the MIT property office to house the project in a rundown wire factory warehouse on the edge of campus. At the abandoned Simplex Wire & Cable Works they shared space with a gigantic wavemaking machine, which the guys from oceanic engineering had installed, and a fleet of dump trucks, which the city of Cambridge had stored for the winter. Way up on a dark mezzanine, whose floor was scratchy with sand from the wave machine, they spread out their wings and went to work. It cost $200 to install fluorescent lights so the construction effort could continue around the clock. On some nights the drafty building let in enough cold air that the five-minute epoxy became half-hour epoxy and humidity from the wavemaking machine made bonding with the water-based glues nearly impossible. But they called it home. "It might be dark and smelly and dirty," John would tell his friends in MacGregor Dorm, "but it's perfect for us because it's just a few blocks from campus and we can't do any more damage to it."

In honor of Kelly Johnson's crew, they renamed Simplex Wire & Cable "the Stinkworks." Production continued apace.

Mark Drela learned a lot about aerodynamics that semester from the boys at the Stinkworks. Like John, he'd come to MIT as a freshman hoping to exchange his hobby of building model airplanes for the secrets of real engineering. He devoured MIT's bedrock theory courses, but continued dabbling every night in his dorm room with hand-launched gliders, the lightest weight models made of balsa and microfilm. Once he joined the Chrysalis team, though, Mark spent all his spare time at the Stinkworks. He slept less than six hours a night so he could stay at the construction site until 2:00 or 3:00 a.m., and often he skipped a string of classes to work on a problem with Parky and Gup and John. At the time, none of them recognized Mark's personal achievements during the construction of the airplane. When he developed a new method for connecting wire fittings on the wings, no one crowed because no one really knew it had been Mark's idea. The Condor's fittings, which they'd examined at the Smithsonian, reached their failure points during tests at 220,000 pounds per square inch. But Mark found a way to coil stranded cable in a thimble, wrap it in copper, and weld it so the connections proved reliable at over 300,000 pounds per square inch. A few parts he built, like the fittings, melded seamlessly to those made by the master craftsmen of the team. He never complained about not getting credit, though, and while he never knew why the older guys worked with such intensity, he never questioned their mission.

"The intent was never really clear," he said. "John would talk about challenging Paul MacCready in the Channel crossing and he'd use words like 'technology readiness' a lot, whatever the hell that means. For Guppy and Parky, I think they just thought it was a neat project. Parky did the fuselage and mechanical parts, Gup gravitated to the model airplane-type work, and John became the de facto project manager. For me, once we started building Chrysalis, everything at MIT just kind of jelled. I thought, Neat! a bunch of guys who like to build."

The project also attracted the skills of a professional engineer at the Charles Stark Draper Laboratory, an advanced space technology research facility in Cambridge. Steve Finberg had spent a good portion of his career at Draper designing digital electronics for data measurement systems and navigation systems for projects ranging from the Trident and Polaris missiles to commercial jetliners. But in his spare time he preferred tinkering with old radios, taking photographs, and rebuilding English sports cars, particularly the Alpine Sunbeam models. Although older than Gup and Parks, Finberg shared their hobbyists' enthusiasm for model airplanes, and after taking pictures of the one-eighth scale model flying in DuPont Gym, he volunteered to build a tachometer and airspeed indicator for them. He toyed with electronic circuits, much like the ones used at Draper, but to place a part on Chrysalis he was forced to cut powers and weights to one tenth their norm. The sensors he designed were unlike anything he'd built before. The challenge of weights and materials opened a new realm of electronics to Finberg. In time, he also learned how to open up the bolted regions of Don Wiener's lab with rest of the guys. Steve Finberg, electronics hacker extraordinaire, joined the weekend and midnight missions into the sanctum of Wiener's shop, too.

By mid-April, while most members of MIT's senior class wrote their résumés and prepared for job interviews, core members of the Chrysalis team scurried to salvage the academic semester. Although Gup had built a computer, which Al Shaw called "the most comprehensive lab project any student has ever done," he never had time to write the required report and it seemed clear he wouldn't graduate. Parky still needed credits for a humanities course and an airplane structures course, which he'd overlooked somehow. Hyong, who couldn't afford not to graduate, vanished from the Stinkworks operation, and even John disappeared for a few days to finish his requirements.

At the time, John's roommate had already been through thirty-two job interviews, and when Lockheed-California announced its interest in coming to the campus, he asked John to sign up with him, to make certain there were enough names to assure the company's appearance. "Sure," John said, "I'll sign up Parky and Gup, too."

They'd assumed the real Skunkworks had halted its operations sometime during the sixties. But as long as they had to show up for the interview, Parky and Gup decided they'd tell the Lockheed personnel agent that the only job they wanted at Lockheed would be at the famed Skunkworks. It was just a joke. Gup would never move to California anyway. Ever since CalTech rejected his undergraduate application in 1969, Gup had been adamant about the West Coast. "Screw the whole state," he said. As for

Parky, he already had an impressive job offer from Boeing, which promised him, as he put it, "significantly the highest starting salary of anyone in the graduating class, if you overlook the minor detail that I'm not going to graduate." But as a favor to John's roommate—or as a kind of joke—Parky and Gup agreed to talk with Lockheed's representatives.

John scheduled them for back-to-back interviews beginning at two-thirty on April 12. The night before, Parky had taken the battery out of his car when he went home from the Stinkworks, so he showed up at the interview wearing grungy jeans and a T-shirt with battery acid holes in it. He also looked a little older than most of the department's aspiring engineers.

"What do you want to do at Lockheed?" the first interviewer asked.

"I want to work at Skunkworks," Parky said. There was a pause. He'd expected at least a chortle.

They looked at Parky's résumé, which included awards from the rocket world championships, a few national championships, his military background, and a human-powered airplane project. They eyed Parky's acid-eaten clothes.

"Why would you want to do that?" a second interviewer asked.

When Gup showed up for the Lockheed interview, he had a large wad of data outputs under his arm. During the past ten hours he'd been working on a modified propeller design for Paul MacCready's Gossamer Albatross team, and he'd nearly forgotten about the time. Over spring break, Professor Larrabee had delivered a paper at the Third International Symposium for Low-Speed and Motorless Flight, and MacCready had taken an interest in the Chrysalis prop. There had been some debate among Chrysalis team members about whether they'd actually give MacCready their own prop design, since they could be considered competitors, after all. But once Parky sensed how much help MacCready needed, he'd relented and Gup offered to run the computer codes to build a new prop for their rival. They hadn't really expected to fly the English channel. But at the very least, Gup hoped Lockheed would grant them a free trip to California so he could deliver the plans. He really wanted to see the Albatross.

When Gup arrived for his appointment, Parky was still in the interview room, talking. When the door finally swung open, Parky didn't walk out. Instead, one of the Lockheed reps poked his head around the door and motioned Gup in. By the time John showed up, Parky and Gup had convinced the Lockheed guys to drive out to the Stinkworks and see Chrysalis. Gup was negotiating for one of them to hand-deliver the propeller design plans to MacCready. That evening the Lockheed interviewers joined John and Gup and Parky for dinner and a tour of Stinkworks. Chrysalis, by then in its latter stages of construction, sparkled in the dingy warehouse. The Lockheed reps couldn't believe their eyes.

Within a week, the personnel director from the company called John and invited all three of them out for a second interview. A week later, they flew into Burbank and were taken on a tour of the L-1011 flight facility and the static test facility where the company examined structural components of the Space Shuttle. After spending the day talking to a long line of Lockheed's project managers and climbing over the scaffolding around the Shuttle, they sat down in the personnel director's office and were given a choice of three of the area's best restaurants for dinner. Then the personnel director said, "Okay, I've gotten a call from Ben Rich, Kelly Johnson's successor, and I've been told to match any offer you've got. We can't tell you what the jobs are, because you'll be working on classified projects. But if you guys want to work at Skunkworks, the door's open."

Parky retained his composure and coolly argued that since he'd been offered a job at Boeing, he wanted to be compensated for the difference in the cost of living between Seattle and Burbank. And that night at the motel, Gup nonchalantly recalculated the loads on Chrysalis's wings, so John took over the task of filling out his application. But Lockheed wanted them, nonetheless. John would do operations analysis, Gup would do aerodynamics research, and Parky would design airplane structures. Skunkworks still operated silently in some black hole of defense department contracts, and there, it seemed, far from the scholastic regulations and bureaucratic morass of MIT, proven engineers—renegades or not—were welcomed.

On June 5, at dawn, a group of Chrysalis team members lifted their HPA out of the Hanscom hangar and carried it down the breezy runway. The airplane looked like a Bauhaus invention in its transparency, size, and revelation of internal structure—it was as big as a greenhouse, as light as a bamboo hut. A few professors who attended the dawn ceremony left their briefcases scattered along the pavement and watched Gup climb into the pilot's seat for the first flight.

"We're going to get a lot of lift now," someone said. Al Shaw aimed the video camera at Gup, as he pedaled and the propeller spun. In the quiet of the morning the team's conversations were the sound of little boys' voices. "Take it to the end of runway five."

Under a cloudy sky, the airplane rolled forward. Gup leaned back, high in the seat, and cranked. Slowly, he rolled, and just as slowly, like a rising patch of fog, Chrysalis lifted up. They'd say later that flying the airplane felt more like swimming than pedaling a bike. It was so stable and quiet. He flew only a few feet off the ground, but a crowd jogged with him.

"Oh wow! . . . That's beautiful! . . . All right! Awright!" Team members and faculty shouted and applauded. The airplane rolled slightly to the left, touched ground, and broke a strut. John screamed, "Shit! The wing broke!" But Gup brought Chrysalis down lightly. "It's okay," Gup said. "It's okay. No damage." Photographers crowded around, and Gup, grabbing the nose, rolled it backwards, and then led a procession to the hangar, first flight completed. After a few tests to improve the roll control, they'd take it back out, again and again, to fly.

Precisely one week later, Paul MacCready's Gossamer Albatross flew twenty-two miles across the English Channel and captured the $200,000 prize for the longest flight by a human-powered airplane in history. The Chrysalis team didn't whine. They'd never been serious competitors. But in the shadow of the Albatross they'd contributed a propeller to MacCready's successful HPA, and they'd built the first human-powered airplane ever to lift off between Lexington and Concord. It was an all-egalitarian show, too. John flew it, Parky flew it, Mark and Steve Finberg flew it. Ed Crawley flew it, Professors Covert and Kerrebrock flew it, Al Shaw and Don Wiener flew it. Even the Albatross's pilot, Bryan Allen, traveled to Boston to fly it. By the end of the summer, more than forty-five pilots had logged three hundred and forty-five flights in the biplane. It flew backwards, it flew in the fog, and whenever a part broke, they had it back in working order within a matter of hours. One day it made twenty-four flights.

The rocketmen had made a flawless transition into a strange new flight regime, and for the first time they commanded a medium that fired people's imaginations about airplanes. Parky liked telling novices that the air inside the wings weighed about half the weight of the wing itself. The truck driver who helped them move the airplane from Stinkworks to Hanscom was awed when Parky told him the air inside the back of his moving van weighed more than the entire airplane. Most people who flew the plane for the first time didn't realize for several moments that they'd lifted off the ground and, stunned, froze at the pedals in midair. Although the team received little attention beyond MIT, John and Gup and Parky achieved a personal victory with their ninety-day wonder. By taking a dare, they'd rekindled enthusiasm for flying machines, even among the faculty, and revived the magical experience of flight for many to share. The aero department awarded them the Admiral DeFlores Prize for the most impressive achievement by undergraduates during 1978–79.

The day before John left Cambridge for California, where he'd begin adult life, he sorted his thoughts about the future once again. This was the final entry in his MIT journal:

> Flight vehicles are the highest embodiment of man's dreams to create, to explore, to move forward. This is embodied both in the high technology required to accomplish missions and in the missions themselves—to carry men to new worlds and open horizons, be they another state, another country or another planet.
>
> At the same time, aerospace is vital to the nation. It is a crucial part of the military and plays an all-important role in the national defense. It's the reason our government is integrally involved in every aspect of the industry. This involvement allows the government to truly shape the industry, so the fortunes of aerospace companies are influenced by the government. Yet almost no one seems to acknowledge this or do anything about it. How are decisions made? Where is the industry going?
>
> You say you favor projects embodying idealism and creative goals, and are saddened that man's most advanced technology is used for destructive purposes. Then how do you rationalize jumping at the chance to go to the most secret institution, the most militaristic of them all? Aside from the fact that Skunkworks historically has made reconnaissance planes, which probably do more to prevent wars than anything else, the answer is the old argument that to be an effective

proponent you must be the best opponent. To debate effectively, you must know both sides. Anybody can blindly oppose the militaristic use of high technology. But only those on the inside know the facts. This will put me inside.

John and Parky were gone by September. The Chrysalis was dismantled, and Gup stayed in Cambridge, hoping to finish his undergraduate requirements and get his degree. Unfortunately, he broke his ankle in a rock-climbing accident in Colorado and withdrew from school. A few months later, he finally abandoned the campus and reluctantly moved to California to join his friends.

LABYRINTH

7

Among the myths of modern engineering, the Lockheed Skunkworks tales must have been among the most exaggerated. At least by the time John and Parky and Gup showed up for duty in 1979, the industry's reputed Olympus had lost its legendary spirit and shuffled along under the weight of government procurement regulations, union rules, and bureaucratic flow charts.

Whereas Kelly Johnson once directed projects with entrepreneurial zeal, enforcing personal standards of excellence among his engineers and machinists, by the time the rocketmen arrived there were committees assigned to every conceivable task. Inspections, standards compliance, program reviews. Most of the great machinists had retired, and many of the engineers who'd been mavericks under Johnson's tenure had been laid off. The firebrands, who would have been managers and mentors to young employees, were supplanted by enervated bureaucrats. "It turned out that the Skunkworks, like everything else, had grown up," John said. "It was just like the rest of the industry."

John spent two years in California with his new wife, Barbara, obtaining the obligatory credential from industry to bolster his vitae. He clocked in and clocked out. He saw engineers go out to lunch every day and waste an hour or two napping in their cars in the afternoons. His boss would leave in the afternoons at exactly five o'clock. The freedom to design and build real airplanes under deadlines mattered less than whether John filled out his timecard properly and had his paperwork available at a moment's notice for audits by government inspectors. He worked on "black" projects—military defense missions—which meant he couldn't even acknowledge the existence of his work among colleagues. He noticed that various levels of security clearances were being used to enhance people's careers, and he saw men collect access codes to information as a means of suppressing ideas or protecting their positions of power. The system was stultifying. In 1981 the Langfords came back east, and John tailored a graduate program combining political science and aeronautics. He opted out of Skunkworks and ended up at MIT. He chose an academic path.

Skunkworks didn't prove much better for Gup Youngren, who'd been assigned to an aerodynamics methods group and spent his time writing 12,000 lines of Fortran code for a new software program called QUADPAN. The program was Lockheed's first attempt to develop a computer program that would analyze subsonic airflow around airplanes, a way to investigate potential structural and stability problems caused by the forces and pressures that dance on the surfaces of an airplane. Although Gup, who had yet to receive a college degree, knew more about computational aerodynamics than the Ph.D.'s in the group, and even though he did receive credit as the software's principal author, he knew that in the long run, engineers with better credentials and more political savvy would outpace him at Lockheed. "It's a big company, and in order to move forward very quickly you have to be quite political," Gup said. "It's sort of like trying to go through molasses." After he wrote the QUADPAN manual, the company turned Gup into a troubleshooter, and moved him around from one temporary desk to another.

Of the three, Bob Parks was the only one who landed a job designing and building airplanes. After spending more than a year at Skunkworks, Parky got a call from the manager of Lockheed's division in Sunnyvale, where the company turned its most creative engineers loose on "blue-sky" projects. At the time, a group from Sunnyvale had latched onto the idea of building a solar-powered airplane and someone had noticed Parky's name in an article about the Chrysalis project. The group's designers knew that, in terms of aerodynamics and structures, an HPA like Chrysalis came as close to matching their needs as any existing airplane. The basic problem was the same—how do you build a strong, ultralight airplane whose power supply is measly and inefficient?

When Parky joined the Sunnyvale team, he not only designed an airplane for them but he also showed them how to hot-wire Styrofoam and introduced model airplane-building techniques that made ultra-lightweight construction possible. "This was very different from what Lockheed engineers—traditionally educated engineers—knew about," said one of his colleagues at Sunnyvale. "You don't follow the usual formulas and come up with an airplane like this. You have to reinvent the formulas. It was pure engineering."

The challenge was to build an airplane that could fly indefinitely at 70,000 feet by drawing its energy from the sun during the day and storing it in rechargeable fuel cells at night. With limited available power, the airplane would have to fly very slowly and float along in an atmosphere above the jet streams. Parky used the Chrysalis airfoil for the design and doubled the length of its wings. When his colleagues balked at how to launch the vehicle—some suggested towing it with helium balloons—Parky assured them he could fly it off an ordinary runway using radio control and a variable-pitch propeller. The Department of Agriculture supported the project because it was thought that such a craft would be invaluable in diagnostics of crops at high altitudes, and NASA provided grants, too, because the applications for communications technologies, such as low-altitude satellites, seemed far-reaching. Theoretically, one airplane might eventually serve as a communications satellite for an entire region of the country.

Parky and four other engineers worked on a feasibility study for a couple of years and produced a two-inch-thick report demonstrating that a solar-powered airplane would be practical—someday. Advances were needed in solar technology, to provide more efficient engines, and the airplane's structural integrity depended on the invention of high-strength plastics that could withstand severe ultraviolet radiation and temperature shifts of up to 200 degrees a day. For a while, their ideas received wide attention inside Lockheed, and then they were forgotten. "One of the big issues," Parky said, "was that you needed a foolproof way to control the airplane so it wouldn't drop down in the atmosphere and get ingested by a 747. That would be real unpopular."

In a few years, after John had returned to school and Gup settled into an office routine, Parky's solar-powered dream withered away. He, too, disappeared back into Lockheed's black hole. Over time, his colleagues at Sunnyvale learned that to keep Parky interested in his work they would have to talk to him about model airplanes or HPAs every once in a while.

As far as Parky was concerned, the ultimate airplanes were not being built in black holes. They weren't being built at all.

In 1984, MIT claimed its first Kremer Prize. Monarch, the fastest human-powered airplane in the world, featured an electronics system so clever and light that the Royal Aeronautical Society impounded the airplane immediately after the flight. The society spent months studying drawings of its intricately designed energy storage system and a boggling cage of circuits that gave the plane speed, climb, and cruise controls, all of which were allowable under its rules but seemed unimaginable to the British judges. The victory also surprised many in the Cambridge aeronautics community, including MIT faculty, because Monarch team members had labored in secret, inverting their sleeping schedules to manufacture parts all night in the basement of Building 33, hiding their handiwork behind cabinets and above their crawl space in Don Wiener's lab. (The evolving mastery of surreptitious building techniques kept them out of Don's sight.) The pilot, it was said, was one of those long-time aeronautics hackers, a flying bum who'd been around the aero department for years without graduating and who wore a T-shirt that said: "Sport Death," an emblem of the brood. One of the team's engineers, an aero department senior named Juan Cruz, had flouted department policy by dropping courses during the airplane construction effort, and eventually left the campus for a job in Kansas without graduating. It was said, too, that the team had help from a couple of MIT types now working in California, who kept track of a rival effort on the West Coast led by a storm of professional engineers. Best of all, the Monarch team had beaten Dr. Paul MacCready, whose Gossamer airplanes had become synonymous with human-powered flight and who had once boasted that he could win the Kremer speed prize by using a bunch of leftover parts from his workshop. After a year of failures and setbacks, MacCready settled for second place and MIT staked top berth in the human-powered universe.

Old Doc Draper's son, James, had heard enough about Monarch to pique his interest. Although his company, Aerodyne Air and Research, Inc., was terrifically busy at the time, Draper rearranged his schedule and fought six o'clock Boston traffic so he could meet Monarch's engineers in Cambridge one October night. The team had just been invited to London to accept the

$33,000 prize offered this time, and the area chapter of the American Institute of Aeronautics and Astronautics had invited them to give an overview of the project. When Draper arrived at the Stratton Student Center, the cocktail and dinner hours had passed and he saw several dozen professors and professional engineers settling in for an evening's entertainment with Monarch's creators. Draper stood in the back of the room while Al Shaw worked a slide projector, and two of the aero department's top graduate students explained the Monarch's technology.

Draper's company was in the midst of intense reorganization and the drive into Cambridge had been frantic, so his mind wandered for the first few minutes of the lecture. But this graduate student, John Langford, told his story with a kind of restrained enthusiasm that was so charming and refreshing that Draper soon found himself engrossed. What was this guy saying? The ideas built into this human-powered airplane were so sophisticated it was almost hard to keep up. The students' enthusiasm for having beaten Paul MacCready seemed almost unnatural.

Monarch, John explained, was part of an evolution of human-powered airplanes, preceded by university-based projects such as BURD and Chrysalis and motivated by a series of Kremer prizes for demonstrations of stability, endurance, and speed. Since 1977, when MacCready's Gossamer Condor won the first Kremer Prize, the airplanes' designs had changed significantly. Engineers were discovering ways to eliminate external wires, improve control systems, use new composite materials such as carbon fiber, and create electrical systems that allowed pilots to store their own energy by pedaling generators that charged batteries prior to takeoff. The new monoplane had cantilevered wings and depended on a greater understanding of low-speed airfoils, a relatively esoteric field of aerodynamics in which Mark Drela, his colleague and former teammate on the Chrysalis project, had become particularly knowledgeable. Mark, whose doctoral work included research about the aerodynamics of low-speed airplanes, was pioneering practical applications for computational theories about these peculiar flight regimes.

The evolution in human-powered flight also promised technological innovations in the area of ultralight aircraft. Advances in the technology, John said, eventually could transform HPA designs into low-powered, high-altitude airplanes that could serve as unmanned platforms for reconnaissance, communication relays, and sampling work in the stratosphere. A winged, unmanned vehicle, an airborne analogue of a lunar rover, would be an effective platform from which to examine the terrain and atmosphere of Mars.

Draper's imagination sparked as slides of different airplanes flashed past. When Mark Drela began to explain Monarch's aerodynamic properties and battery-augmented propulsion system, he sensed an unusually sophisticated technology at work in the guise of an oversized toy. The Monarch team had designed its craft for a flight speed of twenty-one miles per hour, Mark said, just enough to capture the prize for flying around the 1,500-meter course in less than three minutes. Compared to Chrysalis, the wing could be much smaller—175 square feet as opposed to 750 square feet—but the engine had to be far more powerful. Since the rules allowed onboard energy storage, they sought ways the pilot could power up his airplane before crossing the starting line. They rejected stretched rubber storage (a large rubber band, essentially) as too heavy. But a professional engineer near Hanscom, Steve Finberg (also a former Chrysalis teammate), had devised a complex electrical system that worked perfectly. Finberg's circuits transferred the pilot's energy during ten minutes of stationary pedaling into a generator that brought to life a set of discharged batteries; the generator was then converted into a motor by punching a lever with something the pilot called a "Yeager" stick (named for a legendary device used by Chuck Yeager when he broke the sound barrier); and, throughout the flight, the electronics carefully distributed pulses that activated gauges and buzzers to alert the pilot about his airspeed, motor current charge rate, and prop pitch angle. After spending months in the lab testing batteries and soldering circuits, Steve's system weighed only about five pounds and worked with such efficiency that he could predict the batteries would deplete their last volts the instant their airplane crossed the finish line. "There was nothing great about the airframe," Mark said. "It was Chrysalis–Condor–Albatross technology, basically. The energy storage system, on the other hand, did everything but the dishes."

Using a sequence of slides, John outlined a roller-coaster chronology of failures, flights, nose-overs, redesigns, repairs, and piloting errors that almost cost them the prize. After spending the summer of 1983 building the airplane, he said, he, Mark, and Juan Cruz saw many of the early flights end in rubble on the runway. In September 1983, two days after a series of Monarch crashes and repairs, MacCready announced that his speed plane, the Bionic Bat, had won the Kremer Prize. Two months later, when the Royal Aeronautical Society disqualified the Bat because its electrical system did not meet the conditions prescribed by the rules, the MIT team privately redesigned Monarch. In January 1984, John said, he traveled to California to watch MacCready's team as MacCready attempted record flights with his new version of the Bionic Bat. When John returned to Boston, his team started building again, this time in secrecy. Beginning in early April, the Monarch team went on alert every day, starting at 3:00 a.m., to check the weather for clear skies and low winds, to attempt a record flight. While reports dribbled in every few days from California that MacCready's Bat was closer to a record attempt, the Monarch team kept flying. On April 13, they made six flights. On April 23, they made four flights. On May 5, they missed the record by .43 of a second. They tried again on May 7, but the cumulative effects of fatigue grounded their pilot, Frank Scarabino, and then they were stalled for three days while a storm front passed through Boston. On May 11, the team rolled Monarch out of the hangar at 5:00 a.m. and just as the sun was coming up, Frank flew the triangular course in 2 minutes 49.7 seconds, well within the three-minute goal.

When the lecture ended, somebody flipped on the lights and Draper stood, lost in reverie. Good Lord! he thought, these guys aren't building toys. They've made a mechanical device that will jack somebody up twenty or thirty feet and move him horizontally faster than anybody can run. They've tapped the secrets of human flight . . . That's precisely the inspiration Daedalus had when he built wings with feathers and wax . . . The most important expression of flight in history that every aeronautics textbook began with, and somehow these guys. . . .

Draper's memory of the Greek classic was not perfectly clear. He knew Daedalus had been something of a troublemaker, a very idiosyncratic engineer who escaped imprisonment with his son, Icarus. And of course, there was the often-quoted portion of the myth that had made it a cautionary tale, about how Daedalus warned Icarus not to fly too high because the heat of the sun could melt the wax and his wings would fall apart. But something about this Monarch team made the myth resonate in ways other than those regarding the wisdom of fathers and the intemperance of sons. These guys were outstanding engineers, theoreticians, professionals, imaginative zealots—solid MIT stock. He raised his hand.

"I've been sitting back here doing a little wool gathering of my own," he began, "and it seems like this has been an exciting project, but I'm just wondering what are you going to do next?"

"Well, we haven't really had a chance to . . .", John said.

"I mean, you've won the Kremer Prize, but I'm just wondering if there are other options for human-powered flight, like the recreation of the flight of Daedalus from Crete to Greece? To my mind that would be the greatest. . . ."

As Draper talked, John stammered and Mark didn't say a thing. The truth was, they'd been doing calculations for five months and had just received photographs from a member of the team traveling in Europe who had visited Greece to look at possible flight routes. But at this point, only a fool or a braggart could publicly contemplate such a ride. A flight across the Aegean would at least triple the distance flown by any human-powered airplane and pose aerodynamics problems that only the most sophisticated engineers could address. The costs for labor and logistics were staggering. John listened politely and then skirted the issue. After the meeting ended, he and Draper had a long and private talk.

One month after the Monarch Human-Powered Flight Team accepted the Kremer Prize, a proposal showed up in the dean's office at MIT from the department of aeronautics. The myth of Daedalus, it said, "can be turned into a reality." The authors of the proposal had done a cursory search of Greek mythology and proposed flight routes based on accounts of the Daedalus flight by Ovid, Herodotus, Virgil, Horace, Siculus, Diodorus, Sophocles, and Callimachus—distances ranging from sixty-eight to one hundred and seventy miles. At the very least, the shortest route would triple the distance covered by the Gossamer Albatross in its record-setting flight across the English Channel. How could they be so bold? "The answer lies in three parts," the report stated. "The aircraft, the pilot selection, and the meteorology."

With a combination of evolutionary advances in low-turbulence airfoils, advanced avionics, and composite materials, a better airplane could be built. This was the boast. If the airplane could be made so stable that the controls did not require a pilot's constant attention, a world-class athlete could replace an aviator in the cockpit. Taking into account the seven-knot head wind that the Albatross encountered during its flight, simple calculations showed that even MacCready's airplane could have flown nearly twice as far on a perfectly calm day. By amassing weather data and analyzing conditions over the Aegean, precise forecasts could be prepared, and weather would not have such a harrowing influence over the length of a flight.

The idea of recreating the Daedalus flight, the proposal said, represents "a unique synthesis of technology, physiology, meteorology, and classical mythology, and it promises to capture the popular imagination." The report also included phrases like "technology readiness" and "follow-on applications," which gave away the identity of at least one of its authors. Then there was a bold outline of a schedule recommending a two-year project, broken down into three phases: a planning stage; the construction and testing of a prototype; and the redesign, construction, and flight of the final aircraft.

The total budget was estimated at $691,625. An organizational chart showed James Draper as finance chairman, Jack Kerrebrock, the dean of MIT's aeronautics department, as the head of the steering committee, and a working group of young engineers: Juan Cruz, Mark Drela, Steve Finberg, John Langford, Robert Parks, Harold Youngren. Radicals, rocketmen, renegades, hackers.

Every one had his own private reasons for signing up. If undergraduates joined the team as apprentices to the professionals, Jack Kerrebrock thought, the educational value would be enormous. Mark Drela, who was being courted to join the faculty, saw a chance to demonstrate his theories of low-speed aerodynamics. Juan Cruz, already tired of industry's tedium, wanted nothing more than to build a human-powered airplane. James Draper, whose father had developed the navigation system that flew Apollo to the moon, would take his own shot at reviving a mythological dream. And John Langford desired nothing more than to start a business with Mark and Parky, Gup and Finberg, and eventually turn HPA technology into solar-powered airplanes. It was, if nothing else, a mutual yearning for creative freedom.

In any case, they each signed on with the enthusiasm of Icarus himself, the Daedalus myth notwithstanding. Even after MIT and the Smithsonian Institution agreed to provide $74,000 for the group's first year of research, the most vexing antecedent to their plans was disregarded in the initial excitement. They paid no attention to antiquity's infamous cautionary tale. In fact, they referred to themselves, quite innocently, as the Icarus Project.

8

The airplane would be built, first, with customized software. Written originally for commercial applications, those computer programs Icarus Project engineers had once created for giant manufacturers were at last going to serve some purpose other than corporate profits. They'd give lift to a strange new airplane that could fly by the power of imagination in the two-dimensional space of a cathode ray tube.

Out in Wichita, Juan Cruz had written a program that designed graphite parts for the Beech Starship, an executives' transport, at the time the world's largest all-composite commercial airplane. He sometimes referred to it as LASP, Laminate Analysis Software Package. Laminates—synthetic fibers laid out in long, thin sheets impregnated with epoxy—were the most advanced materials made for the engineering avant garde, and Juan's program was one of the few existing tools that made it possible to test laminated structures without getting one's hands sticky in the workshop.

Conceived during Juan's spare time, LASP theoretically allowed Beech Aircraft engineers to select any laminated composite material, such as carbon fibers, build high-performance laminates on a computer screen, and test their strength under any environmental conditions—at the most awful temperatures and at enormous speeds—without having to step into a shop or wind tunnel. With LASP any engineer at Beech could sit down at a terminal and build laminates for wings, tails, fuselages, or spars without risking anything more than the occasional tweak from Juan's interactive program—"Do you feel lucky, punk? . . . Are you sure you want to do that? . . . Go ahead, make my day." The LASP software package, which the project's principal engineer called "probably the best in the world," smart-mouthed bunglers and helped build the sleek Starship ("Real original name, huh?" Juan would crack). Juan used LASP for his job at Beech, but once his share of the Monarch prize money arrived, he bought an Apple Macintosh computer and borrowed ideas from LASP to make new programs that could test HPA structures. His austere little apartment in Wichita became the theoretical testing ground for Icarus Project materials.

Mark Drela created the ISES software for his Ph.D. dissertation, a computational method for analyzing the efficiency of airfoils. ISES (an acronym of esoteric origin) generated geometric shapes for airfoils on a computer screen and predicted their performance. An engineer could build a wing with ISES, modify it, and actually see the results of his work—for example, how an airfoil's drag would be affected by its thickness. MIT faculty members considered it among the best design and analysis codes ever written. Boeing, Lockheed, and McDonnell Douglas bought Mark's program to test their supersonic designs. But Mark used it first, at Mach .03, to plan an escape under human power from the island of Crete.

Lockheed also used Gup's QUADPAN (Quadrilateral Element Panel Method), but because he'd written it as an employee of the company, the software was considered proprietary. Gup finagled a request from MIT to use the program for educational purposes, and when Lockheed assented, he flew it back to Boston and installed it in the aero department. QUADPAN, used in tandem with a Silicon Graphics program, would allow them to test computer models of human-powered airplanes on what looked like a video game screen.

What were sophisticated tools for businesses were toys in their hands. In fact, Juan liked to say that whichever one of them had the most toys when they died would be declared the winner. "In that case," he'd point out, "Parky is way ahead of the pack." It was true. No one ever knew what Parky's programs had generated initially—spy planes, bombers, long-range missiles?—or whether he just wrote fresh ones for fun in his spare time. He could program anything. By the time the Icarus Project began, Parky's home in Silicon Valley worked like Pee Wee's Playhouse, programmed to switch lights on and off at certain hours, to simulate the activity of a household electronically if he and his wife were away for any length of time. A radio alarm buzzed, the bedroom light flipped on, followed by the bathroom light, the kitchen light, a radio—the whole progression of a two-person routine manipulated by what he called "roll-your-own" software programs.

The program Parky wrote for their HPA was one of those "roll-your-own" jobs designed to take all the data from Juan's LASP analyses and Mark's ISES iterations, plug them into equations for weights, aerodynamics, and human power limits, and test literally thousands of different configurations. He tested airplanes whose wings stretched from seventy to one hundred forty feet. He looked at configurations with lots of external wire bracing, like Chrysalis, or with no wires at all, like traditional sailplanes. He flew them at speeds from ten miles an hour, like the Albatross, up to twenty-one miles an hour, like Monarch. This was Parky's "optimization" program. Once Mark and Juan figured the range of their options with the airfoil and structures, they would turn the data over to Parky and he would type it into the program to determine the optimum design. His Hewlett Packard absorbed the numbers like the Looking Glass welcomed Alice, and—higgledy-piggledy—contorted airplane designs into myriad shapes and sizes that in one way or another fit mathematically around a human power plant. "These are what I call rubber airplanes," Parky said. "You can stretch them any way you want."

Computers nudged them toward solutions, but no machine ever helped them ask the right questions. Even with their grand computer programs, creative ideas, and the $74,000 grant that allowed them to fly into Boston or Washington every six or eight weeks and scheme for long weekends together, from the very outset it was apparent they hadn't grasped the complexities of their undertaking at all. They knew very little about the winds and climate of the Aegean. They knew next to nothing about human physiology. They didn't have a notion about how to raise a half million dollars. Worst of all, they were completely unschooled in the nature of myths. Who was Daedalus, anyway? Where did he fly? Who'd pay for it? Their machines were mute.

"A computer is a really stupid device," Juan would say. "Totally stupid."

Arthur Steinberg felt miscast among the working group members when he joined them for dinner at the MIT Faculty Club. There he was, an MIT professor of archaeology, preparing to lecture an assembly of engineers about Daedalus. Across the table sat the former Secretary of the Air Force and top-ranking NASA administrator, Bob Seamans, who'd enthusiastically joined the group as a fundraiser. Dr. Steinberg chatted with professors from the departments of aeronautics and mechanical engineering, technicians from industry. The ironies tickled him. "I can't get over this," he said. He thought of himself as a lapsed and irreverent classicist regaling and teaching engineers about their own peculiar archetypal ancestor in the halls of engineering. No one at dinner really quite understood what Dr. Steinberg meant when he tried to explain that. They just wanted him to tell them which route to take. Some of them, like Professor Ed Crawley and Juan Cruz, listened attentively, and a few others, like Parky and Mark, who really didn't care too much for all the classicist hokum, just barely hung in there to the end. "How far do we have to fly?" was the question they wanted answered. The guy from humanities told them, though a bit obscurely for their tastes.

"Myth making is an accretive process," Dr. Steinberg said, "which causes stories to grow, change, add and delete characters and events, and take on different meanings depending on when they are told. In other words, even ancient myths keep growing and changing every time they get told and used. And now you propose to write another chapter in the Daedalus story by making your own set of wings and flying—but not fleeing—in his trail and hopefully not going too close to the sun—or to the sea. All this for the glory of 'techne,' which is very much in the modern manner of myth making."

Unfortunately, after several hours of searching through classical literature, Steinberg said, he'd found little agreement about where exactly Daedalus flew. It also seemed that the character Icarus was a late addition to the story, an afterthought that emerged in Roman versions during the fourth century B.C. Icarus was rarely mentioned in Greek sources. And, in truth, Daedalus himself proved an unsavory character in almost every version. The Western tradition's first engineer and greatest inventor consistently offended his patrons and invited trouble into his life. This much was certain: Daedalus was an ingenious engineer, true enough, but also quite a scoundrel.

Other than the god Hephaestus, Daedalus was the only major character in Greek mythology who represented the essence of engineering, particularly in terms of cleverness and craft. He was a mortal, though, not a god, instructed by Athene in the royal house of Athens. Century after century, storytellers attributed the development of essential tools and jigs to his hand. He invented the saw, ax, plumbline, gimlet, glue, the ship's mast and rigging, plumbing, and the labyrinth. As an artist, he's said to have been the first to sculpt figures in lifelike postures, underscoring the importance of movement and spirit in human subjects rather than rendering them rigid and gestureless. But as a man, he resembled any other man and, as such, was sometimes slave to petty jealousies and a mortal's torments.

Daedalus, Dr. Steinberg began, had a nephew who shared his gift for engineering. Although the boy, Talos, worked as an apprentice to the engineer, by the time he was twelve years old Talos' accomplishments rivaled Daedalus' own. The boy forged iron in the shape of a fish spine and produced an excellent saw. He invented the potter's wheel and a compass for making circles. As the boy's reputation spread in Athens, Daedalus became enormously jealous. One day he invited Talos up to the roof of Athene's temple at the Acropolis and, pointing out features of the landscape, he walked the boy over to the edge and shoved him off. Daedalus rushed down to the steps of the Acropolis, bundled the corpse into a sack, and stole off to bury it outside the city. But there were witnesses who saw him hauling the blood-stained bag. The great engineer was charged with murder.

Accounts varied about whether Athens' rulers banished him or whether Daedalus escaped before trial. He disappeared from Athens and reemerged in Crete, working at a new job in the palace of King Minos, a son of Zeus. Minos ruled a peaceful civilization from his estate, but those who knew him considered him suspicious, selfish, and untrustworthy. It was said that he'd even deceived a god. Years before Daedalus arrived, when Minos claimed his position as king, he'd promised to sacrifice a bull to the god of the sea, Poseidon. But when the ocean surrendered a magnificent white bull for slaughter, Minos ordered it hidden among his herd and sacrificed a lesser animal in its place.

When the king married, Poseidon sought revenge for the affront, and instilled in the king's wife, Pasiphae, an unnatural passion for the white bull. While Minos enjoyed many affairs with slaves in the palace, his wife lusted secretly for her husband's most prized animal. In time, she learned she could confide only in Daedalus, who seemed less disturbed by wicked desires than most men. Slowly she devised a plan for him to help relieve her of the craving. Daedalus became Pasiphae's conspirator in the rush to satisfy her sexual desire.

Daedalus sculpted a hollow wooden cow for Pasiphae, stretching cowhide over the device, concealing wheels in its hooves, and making doors for its back. On a pre-assigned day, Pasiphae crawled inside and thrust her legs down its hindquarters. Daedalus rolled her out into a meadow near Gortys, where the white bull grazed. The engineer discreetly slipped away, and Poseidon's bull mounted her, plunged into the wooden creature, and impregnated the queen. Their union produced a monstrous child with a bull's head and a human's body, an evil creature called the Minotaur. In his horror, Minos consulted an oracle for advice about how to hide their shame. The oracle told Minos to order Daedalus to build a labyrinth at Knossos, a maze from which no one could escape, and to isolate the monster at its center. Some time after Daedalus finished making the labyrinth, Minos discovered that his chief engineer had been part of the queen's deception, and forced Daedalus into his own maze, to face the Minotaur, to suffer and to die.

The lurid details about Pasiphae's lust and the conception of the Minotaur had never appeared in modern aeronautical texts. The story sounded far different from the moralizing tale taught in grade schools. The engineers around the table were entertained by Dr. Steinberg's revelations. The story of the escape, however, was more familiar, and the engineers patiently listened to the retelling of how, during the day, Daedalus gathered shedding feathers from birds that flew over the maze, and constructed wings using wax from the candles that lit its infinite corridors at night.

According to the most popular version, by Ovid, Daedalus also built wings for his son, Icarus, whom Minos exiled, too. Before their escape, the father carefully warned his boy not to fly too high, lest the sun melt the wax, and not to fly too low, because the ocean's moisture would waterlog the feathers. No mortal had ever flown before, and as the two rose up under their own power, Icarus was increasingly excited by the flight. They passed the island of Samos, by Delos and Paros, and as they flew, the son wandered higher and higher. Despite his father's warnings, Icarus darted beyond their route, arched toward the sun; the wax melted and the wings came undone.

The path of Daedalus was uncertain. Some said he snatched his son's body out of the Aegean and flew to the nearest island to bury him. Some said he flew east toward Turkey and buried the boy on what is now the island of Ikaria. Other stories suggested Daedalus redirected his course northwest to the Peloponnesus, the Greek mainland, and from there made his way to Naples, where he dedicated his wings to Apollo.

As Dr. Steinberg spoke, John Langford looked at his maps and traced with his finger the different routes from Crete to Ikaria, from Crete to the Peloponnesus, from Crete to Naples. He glanced skeptically at their guest.

Despite uncertainties about his flight path, Dr. Steinberg continued, Daedalus reappeared next in Sicily, where King Cocalus hired him as chief engineer for the island kingdom. As on Crete, Daedalus made toys to entertain the king's family and improved upon his Cretan plumbing systems to create buildings with hot running water. He made particularly good friends with the king's daughters.

In the meantime, King Minos, enraged by Daedalus' escape, left his palace to search the Mediterranean for his miscreant engineer. He went from palace to palace carrying a puzzle he knew only Daedalus could solve. In Sicily, Minos encountered King Cocalus, handed him a shell and a piece of string, and asked him if he could find a way to thread the string through the object. Cocalus privately passed the puzzle along to his engineer. Daedalus tied the string to an ant, bored a hole in the end of the shell, dabbed honey on the hole, and let the insect course its way through the maze toward the lure. With the twine sufficiently wound inside, Daedalus merely pulled the thread through. When the puzzle was returned, solved, Minos demanded Daedalus' surrender. While King Cocalus' daughters stalled Minos, Daedalus reconfigured the plumbing in the guest quarters of the palace, and one night when Minos went to relax in a warm bath, they flushed boiling water through the pipes. Minos died in the scalding tub. A war ensued between Sicily and Crete, and Daedalus escaped again, perhaps to Sardinia, where artwork called Daedaleia survives even today.

"The trials and tribulations of an engineer!" Dr. Steinberg exclaimed. Despite the contributions he made to advance civilizations, Daedalus was forced to serve many masters. Like any other engineer during the Roman and Greek renaissance, he was viewed with contempt because he worked with his hands. Plato, Plutarch, even Aristotle had been outspoken in their disregard for engineers. In fact, craftsmen were often slaves, freedmen, or foreigners, but rarely were they citizens.

John applauded along with the rest of those around the table. Dr. Steinberg, he thought, had done an adequate job of retelling the myth. But he hadn't told them much more than that. He looked at his maps. You can't depend on an MIT professor for something like this, he thought. We should have invited someone from Harvard. This guy didn't even pass out a reading list.

After the speech, the working group broke up, and then a few members privately convened to rehash Dr. Steinberg's comments. If the flight of Daedalus had spawned so many paths, then perhaps they could simply choose the one that suited them best. Rather than worry about a literal duplication of the mythic flight, they decided to defer route selection and focus on building the best advanced-technology aircraft possible. Juan and Parky spoke to Kerrebrock about returning to MIT so they could finish their undergraduate requirements and build the airplane. Steve Finberg offered to design a set of small weather collection stations that could be anchored in the Aegean for a year. A young professor of aeronautics, Steve Bussolari, volunteered to do meteorological studies and begin recruiting athletes to pilot the airplane. John made a list of possible sponsors, including Union Carbide, IBM, Hewlett Packard, and DuPont.

They met until after midnight Friday, began again at 9:30 a.m. on Saturday in Building 33, adjourned fourteen hours later, and reconvened in the aero department library at midnight. On Sunday, they met again at 10:00 a.m. and worked until late afternoon, when Parky and Gup left to catch a plane for California, Juan caught a flight back to Wichita, and John, whose dissertation was now being funded by the Institute for Defense Analyses in Virginia, left for his new home in Washington.

The story of Daedalus receded in their minds. Dr. Steinberg never returned to the working group sessions. At some point, perhaps, they could find someone at Harvard or Yale to help them explore the many routes that Daedalus might have flown. Until then, they faced the toughest technical challenges of their lives. With the aid of advanced software, corporate funding, and unbounded imagination, they'd overtake the Daedalus route with the first human flight from Crete in thirty-five hundred years. Interestingly, even after the classics lecture, some members of the team continued to refer to their enterprise as the Icarus Project. References to escapes or mazes were rare.

9

The working group's meetings broke out into intellectual dogfights, paper airplane battles that stretched from twelve to fourteen hours a day. Some people copied formal position papers and others scribbled on a chalkboard or napkins during their weekend presentations. Juan Cruz—never one to demur in the face of outright stupidity—sometimes offered a one-word critique to his friends' worst ideas. His recurring gasp characterized the tenor of the meetings. "Woof!" Juan would say, quietly but firmly. A pack of baying hounds couldn't raise such a scare.

Parky and Gup carried on a kind of Socratic dialogue that couldn't always be distinguished from capriciousness. They tangled over extreme ideas about hinged props, planes with two tails, fuselages that looked like pods, and launch sites 8,000 feet above sea level. They'd both talk at once, neither giving in, until after a half hour or more it would become apparent that Gup and Parky had switched positions. John recognized the behavior from their rocketry days, when he'd hear his friends haggling over parts—the seller would offer a price, the buyer would offer a lower price. But instead of splitting the difference, they'd diverge, the seller asking for more, the buyer demanding even less. They were incorrigible.

"That's enough!" John would yell. "We're two hours behind schedule. We need to move on!"

"Back and forth, up and down, side to side," Juan would say. "Woof."

Guests from the academy rarely made it to the technical sessions on Saturdays. Instead, they joined the team at Friday night dinners, found themselves fêted, then questioned incessantly afterwards. The crowd there wasn't so querulous, but the level of debate remained rich. Scholars would come in, amused at crossing over from their separate disciplines to address a group of engineers. Sooner or later they heard the challenge. An early guest, Dr. Ethan Nadel, a senior physiologist from the John B. Pierce Foundation at Yale University, preceded his talk by distributing copies of an article he'd just written for *American Scientist* about endurance exercise, then gave a generic physiology lecture. He hadn't expected much response. The working group parried with detailed questions about glycogen storage, anaerobic processes, oxidation, and the cooling metabolism. "Our data show that we need a pilot who can produce two hundred fifty watts for four hours," John said. "But we haven't found any evidence in the literature for power levels at these durations. From your experience, what can you tell us about . . . ?" After an hour of this, Ethan decided to join the group on Saturday, too. After another three hours of frank exchanges over the limits of human endurance and possibilities for testing elite athletes in Greece and the United States, Ethan realized they were serious. He signed on as the team's physiologist. "The best ideas are the most unorthodox ideas," Ethan said. "This is venturing into the weird. This is uncharted territory."

The Smithsonian assigned a young historian from Wellesley to the working group. Debbie Douglas would take the "Drela Express" in from Logan Airport on Friday night expecting a gray, two-day drizzle of calculus and statistics. Instead, she found papers plopped in her lap with peculiar letterheads—Stagnant Technologies: Paralysis Through Analysis; Reflective Technologies: We Do It With Mirrors; Traditional Engineering, Inc.: Where Yesterday's Technology Is Good Enough; Aerostuff: If the Big Mac Does It, We Do It Better. She attended their show-and-tell hours, when Parky demonstrated his latest toy (a portable Hewlett Packard computer in a suitcase, which they called a "luggable"). She watched Juan pass out samples of strong, light plastic sheeting called Mylar, all kinds of plastic foams, and graphite. She even joined them in a visit to the Greek Embassy in Washington to make their first formal request for foreign assistance, at the least propitious time—they met just days after terrorists hijacked a jetliner from Athens and President Reagan angered the Greeks with televised warnings to Americans to stay out of Athens.

Parky would snipe that Monarch had been "a boring airplane," and Juan would groan. Juan would promote the most difficult flight path, insisting that in the best known version of the myth, Daedalus flew to Ikaria. Every once in a while he'd say, "Let's fly it to Ikaria! One hundred and fifty miles in eight hours!" He was ignored. And Mark, who generally kept quiet, would occasionally walk up to the blackboard to update his latest applications of low-speed aerodynamics. The group would fall silent. When the time came to talk about designing low-speed airfoils, nobody could seriously question Mark's expertise. He'd bypassed plenty of professors in that realm already. "Mark's stuff was wild," Juan said, "and we were totally clueless."

John found himself, once again, managing a reckless team of radical engineers. He backed Gup in arguments with Juan over the wisdom of building an autopilot so the airplane could fly at night. Gup argued for advanced avionics; Juan considered it unnecessary baggage, "whiz-bang technology." And when Parky challenged Juan about his spar designs, or taunted, "Verify your numbers," John stood clear.

It was like old times. The sparring matches weren't always gentlemanly, but everyone acknowledged the scientific method and acquiesced to truth, as best they could judge it. No single member of the working group could claim solitary rights to an idea. Juan was the structures expert, but he listened to every word cf Parky's critiques. Mark would shape the wings, but Gup offered ideas that could change the shape of the entire airplane. Steve Finberg understood electronic devices in an almost mystical fashion, but no one withheld comments about his proposals, either.

By the fall of 1985, when it seemed clear they would soon settle on a design, John wrote every member of the team a letter, stressing that an egalitarian nature had always been their source of strength. And while he had assumed the role of "de facto project manager" he would see that they worked, as much as possible, without hierarchy and without bureaucratic interference. Juan quit his job at Beech. Parky made plans to leave Lockheed. Under the sway of imagination and the influence of intellect, anything seemed possible.

Juan's stomping ground at MIT had been underground. Frightfully thin, with wild, rangy red hair and a stinging wit, he looked more like a bohemian poet than an avant-garde engineer. The resemblance, as it turned out, wasn't just coincidental. More than anyone else, Juan lived a radical existence—sleeping during the day, working all night; turning down job offers from any aeronautical firms with ties to the defense establishment. Wherever he lived, he lived simply and passionately. He would sleep on the floor of his apartments, tend a single shelf of revered aeronautical classics, listen to momentous jazz recordings, and devote himself to the creation of the world's lightest materials with which to build airplanes. He was fascinated by the properties of composite materials, particularly the uses of graphite epoxy.

He'd spent most of his years as an undergraduate in the composites laboratory, building and exploding graphite epoxy tubes. For two years he worked as an apprentice to a graduate student, Michael Graves, whose doctoral thesis, "The Catastrophic Failure of Graphite Epoxy Cylinders," had become a six-year mission. In an oppressively dark concrete blast chamber, Juan made carbon tubes—a foot in diameter, two feet long, 1.5 millimeters thick—pressurized the cylinders with nitrogen up to one hundred seventy pounds per square inch, and dropped blades from a miniature guillotine on them. The explosions shook professorial lairs several floors up. Often Juan worked at night—all night—and regularly skipped classes so he could learn by feel the properties of graphite. He missed meals so frequently that Graves worried about Juan's health, and learned to keep him alert with a steady diet of coffee, sugar, and frozen M&Ms. Over a period of months his hair grew long, and he developed an ironic sense of humor that helped him endure at MIT—but just barely. One year, during an academic slump, he dropped out to work for Sikorsky Helicopters in Connecticut, building composite specimen samples. A couple of years later, when the time came to graduate, he dumped two humanities courses at Harvard so he could help build Monarch. Nothing mattered as much to Juan as building a human-powered airplane.

Professors generally liked Juan—he went to Beech highly recommended because of his expertise with graphite composites—but they did not really understand him. "Aerodynamics is normally very abstract," Juan would say. "Structures is very physical." The palpable nature of engineering appealed to Juan—he liked building pieces of airplane frames with his hands more than noodling over obscure theories. From the time he was a child in Puerto Rico and read *The Spirit of St. Louis*, Juan had identified with the hangar engineers who took notes from their pilot and then built a machine for a specific mission. Even when he worked at Beech, his bosses learned that if Juan wasn't in the office, where all the other engineers worked in pressed white shirts and plaid ties, he could be found in the shop, perhaps shirtless, perhaps shoeless, clambering into the belly of a new fuselage or slipping through the exhaust vents of a jet engine, looking at the workmanship, examining the craft.

Mark Drela's place in MIT's institutional maze was much higher than Juan's. In his well-ordered office in Building 37, Mark could let his eyes wander out a window framing the landscape between MIT and Vassar Street. His shelves held heavy books on aerodynamics theory, a few model gliders with fifteen-inch spans and microfilm wings twelve millionths of an inch thick, long-endurance gliders weighing between eight and ten thousandths of an ounce. Since his days as an undergraduate, he'd built some of the best lightweight gliders in the world. But his reputation among the faculty rested on his propensity to advance aerodynamic theory. Professor Larrabee called Mark "the Chopin of applied aerodynamics." He spent hours and hours at a computer terminal, quietly honing new ideas.

Late at night, Juan would wander in.

"How ya doin', snake?"

"Juan?"

"The Juan and only."

"Yes?"

"Getting your regular dose of CRT radiation?"

"Juan, what do you want?"

"Say, does that thing work on the abacus principle?"

Although Mark had made a conscious effort to leap MIT's academic hurdles and Juan made a conscious effort to avoid them, each admired the other's uncompromising attitude. Maybe Mark stood on the top academic rung, but Juan knew he didn't get there by playing politics. "Does he look like a 'Yes' man?" Juan would say. "A 'Maybe' man, an 'I'll get back to you' man. But not a 'Yes' man." And Mark appreciated Juan's humor and engineering savvy. Despite their contrasting ranks in the academy, they were craftsmen, after all, with a common purpose. Mark's theories would shape the wings, Juan's graphite would form the bones.

* * *

The material for bonemaking came in long black rolls packed in dry ice. Unfurled, the sheets showed thousands of minuscule synthetic fibers, running in long parallel lines like threads in strapping tape, all embedded in a gooey epoxy adhesive. Pliable, flimsy, easily shred, carbon fiber buckled readily to pressure applied perpendicular to the sheets. Its most amazing property, its strength, lay along the direction of the fibers. Over the years, Juan had learned to take advantage of graphite's qualities by wrapping the sheets at specific angles around a firm shaft, called a mandrel. Layers of crisscrossing sheets, wrapped at angles from zero to 90 degrees, formed a bed of material that could be as strong as steel and far, far lighter.

A natural analogy to composites would be body tissue, which has high strength and flexibility. Juan's favorite analogy, though, was the Coke can. The lightweight cans that anybody can crush by squeezing from the sides will bear an enormous weight from the top or bottom—Juan could stand on top of one without crushing it. The same principles applied to making airplane structures. A decent aeronautical engineer could tailor carbon fibers to handle air pressures and forces of gravity that might cause an airplane to bend at a certain point or twist a certain way along a wing or on the fuselage. A good engineer could calculate those so-called bending moments and torsional effects well enough to save an enormous amount of weight. And a great engineer could figure out the specific forces on one airplane so well that if any single part broke, the whole airplane would crumple. This was Juan's goal: "If you build the tail boom so it won't take loads at over thirty miles an hour, and you fly over thirty miles an hour, the tail boom should break, and the wing should break. It should all break at once. Overdesign adds weight."

At Beech, as in most of the industry, engineers used graphite in place of aluminum. A graphite epoxy tube weighs about half as much as aluminum, can be up to twice as stiff, and four to five times stronger. But rather than designing pieces to withstand the particular aerodynamic forces they'd bear, most engineers simply transferred their knowledge of metals and built parts to simulate aluminum, laying graphite fibers to give their parts the same stiffness in all directions, just like a metal. It wasn't a wasted effort, but it wasn't efficient, either. At some companies, Juan saw engineers actually riveting graphite pieces together, just as they'd do with aluminum, rather than gluing them. "Barfo," Juan would say. "If you want to put a hole in something, put it in metal." The advantage of customizing their own materials was lost on traditional thinkers. The tripe they produced was called "black aluminum," and Juan wouldn't go near the stuff.

Black aluminum, like its metallic counterpart, was isotropic. In other words, its stiffness and strength were the same in every direction and, consequently, would not necessarily yield the most efficient parts. On the Daedalus airplane, Juan could predict the basic pitching moments, torsional effects, and shearing forces along the airplane's wings, and design graphite fibers at each section of the spar to absorb those particular pressures of flight, but not much more. The benefits in efficiency and weight savings were critical to making the Daedalus flight a realistic proposal.

Most of all, Juan wanted lightweight bones. A single kilogram would increase the pilot's power by nearly 2 percent, and since the largest percentage of weight would be in the wing spars and tail boom, Juan had to build lean tubes with as few layers of graphite as possible. The walls of the hollow tubes that comprised the spars would have to be as thin as a dime. He liked to describe his job as "adding lightness."

Like Juan, Mark had learned his craft by experimenting in workshops. Only his first shop wasn't in the basement of MIT, but in the basement of a modest house in Kedzierzyn, Poland, where he was born. "I was always asking my parents and grandparents, how do airplanes work, how do they fly, what makes them go?" he said. When he was three years old his parents gave him a toolbox with a mallet and nails for Christmas and he went right to work in the basement of their home. One day his mother heard a commotion downstairs and when she ran down the steps, she found Mark hovered over a pile of wood. "He had three boards," she recalled. "A long one and a shorter one and a still shorter one. I looked close and realized this was an airplane, and I said, 'Mark, what are you doing?' And he said, 'Mommy, I'm building a plane.' I thought, Fine, let him play with his mallet and little nails. Well, a few days later Mark came up crying and he said, 'Mom, the plane is so big I can't get it out the door.' So his dad helped him take it apart and put it together again in the backyard. He was very interested in models after that."

By the time he was five, Mark could read the instructions needed to build plastic models. By the time he was eight, he'd made an arrangement with the owner of a kiosk in their neighborhood—a little newsstand that sold plastic airplane kits. The kiosk's owners would give Mark kits for free, and he'd build models for displays around their store. He took so much pride in his models that he'd place them on his bed at night and he'd sleep on the floor.

As a teenager, after his family moved to the United States, Mark discovered balsa. In Poland, the most readily available material for model builders was spruce. But by using the lighter, more flexible balsa woods in the United States, Mark improved his skills and started competing against the nation's best aero modelers. At seventeen, he'd set four national endurance records for lightweight model gliders, rubber-band-type airplanes that would fly, at walking speed, for up to fifteen minutes under their own power. Also during high school, he built his own wind tunnel using a large electric fan, and tested tiny airfoils made of Venetian blinds, cardboard, balsa, or aluminum from thin-walled Camembert cheese cans. He cared little for gasoline-powered models—"The noise, the mess, the smell, the expense turned me off," he said. Instead, he zeroed in on ways to better harmonize wings with air. "It's almost an aesthetic thing," he said. "You learn to do the most with what you have."

Although even as an undergraduate at MIT his understanding of airplanes surpassed most of his classmates', it wasn't until he entered graduate school that he began to grasp the aerodynamics of small models. Physical reactions that occurred on the wings of models differed significantly from the physics of ordinary airplanes. Their diminished size, slower speeds, and the minuscule span and width of the wings placed models in a class more like birds than 747s. It all had to do with an abstract formula—combining wing size, speed, density, the viscosity of the air—which yielded a figure called a Reynolds Number. Basically, ordinary airplanes flew at Reynolds Numbers over 1 million, and models, birds, butterflies—and human-powered airplanes—flew at Reynolds Numbers less than 1 million. Not much was known about low Reynolds Numbers. It was a murky, unexplored domain.

A Reynolds Number tells something about how air behaves as it moves around a wing. Like rain or misapplied oil from a mechanic's oilcan, air runs over a wing like any other fluid, washing across but leaving a thin film on the surface. It sticks. Air happens to be invisible to the eye, but it is sticky nonetheless. Mark explained the notion of the air's stickiness, or viscosity, like this: "If you take two plates and wipe their surfaces with honey, oil, or plain air, then put them together, face to face, and slide them past each other, it takes a different amount of force to move past each fluid. If you have honey, it's hard to slide the plates past each other. With oil, it's easy. With air, it's almost nothing—shwwwt! you go right over it. So viscosity is what tells you how hard it will be for the fluid to perform against a surface."

The size and speed of an object also make a difference in how it moves through air. Consider a bacterium going through a room, for instance. To a bacterium struggling across a room, the air feels like honey. The organism is so small and so slow that the air feels very viscous. Now, consider your hand moving through a bucket of honey—in scale, it's comparable to the bacterium struggling through a room. Take it a step further, and imagine the strain of an ocean liner cutting through hot tar. In each of these cases, the Reynolds Number may be the same, though the viscosity of the fluid is drastically different.

A Reynolds Number of 1 is quite small. A bacterium in a room or a dust particle floating through air has a Reynolds Number that low. So would your hand, as it moves through honey, or an ocean liner, if it battled through a sea of tar. But if you took your hand out of the honey bucket and waved it through the air, the Reynolds Number would change. A hand waved in the air yields a Reynolds Number of about 10,000. In other words, the air feels one ten-thousandth less viscous to your hand than when it moved through honey. In a 747, because the airplane's speed and size make its movement through air almost effortless, the Reynolds Numbers are in the hundreds of millions. The viscous effects are insignificant.

Mark's devotion to model airplanes, particularly the lightweight gliders, drew him into this little understood region of flight. Compared to traditional aerodynamics, theories of low Reynolds Numbers could not be tested well in wind tunnels. The air that whisked through wind tunnels made little difference when testing a jet's wings, but because of the slight, nearly imperceptible turbulence created by a rush of air, wind tunnels couldn't accurately simulate the forces that caressed the wings of birds, dragonflies, and model airplanes. Slow, lightweight, gentle creatures tested best in the airless brains of a mainframe computer. Making computational models for such wings was Mark's genius.

Understanding how an airplane flies is relatively simple. It's basically a matter of seeing how the airfoil manipulates air as it flows over the wing.

The earliest pioneers of flight in America learned about air by watching winds. In the nineteenth century, for example, Octave Chanute was laying railroad track in Illinois and Kansas, and from time to time he'd see high winds that lifted roofs off houses, swept locomotives off their tracks, and tossed bridges from their platforms. These dramatic events weren't attributable to the velocity of the winds alone, Chanute could see. Somehow trains, bridges, and roofs weren't being bowled over by the impact of air, they were actually lifted up by the flow of air around objects. There was a strange principle at work, a significant force playing around the shape of things.

In those days, it was reasonable to expect engineers to dabble in a spectrum of scientific principles, which would lead them from building bridges to laying railroad tracks to designing boats. Chanute's observations about lift caused him to shift his attention from railroads to flying objects. Like others who shared his peculiar interests, Chanute worked in isolation on the puzzle of lift and flight. It's said that in 1891, when he'd stopped in Kansas City to dine with a friend's family, his hosts asked him about his hobbies. Embarrassed, he replied, "Wait until your children are not present, for they would laugh at me." Chanute's observations eventually yielded what were probably the first scientific evaluations of glider flights in America and the data he accumulated on control and equilibrium contributed to the success of the Wright Brothers' flights at Kitty Hawk.

Flight's pioneers learned that air not only acts like a fluid as it washes over objects, but it has mass and pressure as well. Heavy winds snap tree branches; air blown into a bag expands the sides. Although its force might not be apparent, the air pressure weighing on this page, for example, amounts to between six and seven hundred pounds. The molecules that form air are also so densely packed that each one collides against another five thousand million times per second. But as impressive as these figures sound, and as ferocious as they can be when bundled up in the power of a tornado, for the most part, air can be easily manipulated.

When a curved plate (like a Frisbee) is hurled through the air, the plate makes a continuous slice. A rush of air meets the leading edge, and some molecules slip up and over the top surface and the rest dip down and course across the bottom. This drastic cut creates a flow of air above and below the plate; two distinctly different air pressures form. Over the top surface, where the molecules accelerate upward, the centrifugal force creates a suction on the plate, and air pressure drops. On the bottom surface, where the molecules are displaced downward, air pressure increases. If the bottom surface is slightly concave, the molecules driving across the bottom are fooled into thinking they're going to rise, and the centrifugal force pushes up. Depending on the contours of the plate and the rate of speed at which the air moves across the surfaces, this differential between pressures will cause the plate to rise or float. Very simply, that's how the wing of an airplane works. In practice, the manipulation of air around a contour—an airfoil—can be turned into a significant anti-gravity force.

Looking closer at air as it flows over a wing, scientists have found all kinds of action taking place across the top surface. The air flows across the wing all right, but near the surface there is a considerable amount of friction. Air, like water, oil, and honey, clings to all surfaces. And when it clings to a wing, it creates a force called drag. Drag on an airplane, like weight, requires more power from the engine to compensate for its sluggishness.

Obviously, for an airplane engineer, the idea is to shape wings that promise the most lift and the least drag. One way, of course, is to make certain the surfaces are clean and smooth, to let the air glide with as few disturbances as possible—no grit, no grime, no ice. Another way is to streamline the shape of the wing, so the air is compelled over the top of the contour and the rising molecules are allowed to drop gently along a subtly sloping downward gradient. A sharp edge along the back of the wing—the trailing edge—serves the important function of guiding the flow so the rushes of air above and below the wing meet in an elegant confluence at the end, without creating eddies of air or turbulence that could disturb the flow in its wake. A mass of eddies or burbles—a jumble of air molecules next to the surface—is called turbulent flow. Turbulence, obviously, causes drag. On the other hand, air that's made to accede to the wishes of the wing, that flows without any disturbances, is called laminar flow. In most cases, even on very good wings, the air begins with a laminar flow and as it moves rearward it undergoes a transition and turns turbulent. In any case, engineers look for shapes with a high ratio of lift to drag, to minimize turbulence as much as possible.

For making an ordinary airplane fly, the shape of the airfoil does not require critical analysis. Anyone with an aesthetic sense could draw a passable airfoil freehand. Make the leading edge slightly plump and rounded, arch the top to a slight peak, and let it slope gently down to a sharp trailing edge. Make the bottom run straight or arch it just a bit. That'll probably do the trick.

But at low Reynolds Numbers the physical nature of air, as it flows slowly over the wing, makes the airfoil's design crucial. This was Mark's problem in designing the wings for the Daedalus flight. Because the human engine could only produce a small amount of power and even the shortest Daedalus route under consideration would be at least sixty-eight miles, the airplane would necessarily fly slowly, somewhere around fifteen miles an hour. Taking into account the airplane's size and speed and the viscosity of the air, Mark knew he'd be working with a range of low Reynolds Numbers—between 200,000 and 600,000—about the same as the flight characteristics of a remote-control model airplane.

The most critical aspect of the wing's performance depended on a thin stream of air, ranging from about a hair's breadth at the leading edge to the height of a penny at the trailing edge. It's called the boundary layer. Mark's goal was to guide that little ribbon of air around the upper surface as far as possible without letting it switch from smooth laminar flow to draggy turbulent flow. Think of the boundary layer as the cars on a roller coaster—Mark needed to design the slope on his roller coaster so there'd be enough momentum at the leading edge to carry the cars over the biggest hump, but without such a jolt that the passengers would scream or quake or want to bail out along the way to the summit.

Unfortunately, there was a real problem with boundary layers that travel over slow-flying wings. They tended to be so perfectly laminar that they became lethargic, lazy, and finally worthless when it was time to produce lift.

Mark's design had to invite laminar flow over at least the first 40 percent of the wing. Like cars hurtling over the first hill of a roller coaster, the airflow had to be fast enough to carry them to the top of the biggest hill. Without a good rate of speed, there wouldn't be enough momentum to carry the boundary layer over the most important hump. If the layer lost its momentum, it would do just what the cars on a wimpy roller coaster do. It would slow down and stop. But instead of rolling backwards, like the roller-coaster cars, the layer would simply peel off the wing. It would bubble up off the surface and float away. At such times, the effects of drag hamper performance. An airplane will quit flying. The wing stalls.

Ironically, there was only one major force that could keep the boundary layer from separating permanently from a low-speed wing: turbulence. If Mark could shape the airfoil to get a maximum amount of laminar flow and then tweak it at just the instant the boundary layer separated, he could create enough turbulence to assure that the layer would reattach to the wing. Again, using a roller-coaster analogy, Mark's job was to keep the cars moving gracefully toward the peak—not going so slowly that there wouldn't be enough momentum to reach the top, nor so fast that the ride would drive the passengers into a frenzy on the way up. The roller coaster would have to be made so that at the instant the first car languished at the pinnacle, the passengers would have to be startled enough so their own jolts and twitches and screams would drive the lead car over the peak and let gravity draw them down the back slope. In essence, at just the right moment, Mark had to play a trick on the air: He had to create turbulence.

Tapping on the keys of a Perkin-Elmer 3242 computer, Mark plugged all his calculations into the ISES program and let the mainframe figure out the contours for this crazy roller-coaster ride. Staring at the glowing green screen, he would watch his numbers vanish, and after a few minutes' pause, there'd be a blink, and then a quick sketch would appear in bright, luminescent green. So this, he'd think, is the ideal airfoil. With the flick of a finger, he'd pop a key on his control board and the screen would blink again and open a little window, so he could zero in on a specific section of the airfoil to examine the theoretical boundary layer. Something was wrong.

The problem, he discovered, was that the ideal airfoil was simply too perfect. In the airless brain of a computer, the wing didn't have to contend with any surface imperfections. There was no such thing as a fly speck on the ideal airfoil or a plastic wrinkle or a hint of dew. The ideal airfoil was a risky design, keeping the boundary layer right on the verge of separation at every moment. In an airless space, it was no problem. But in reality, any slight perturbation against the surface—even the sound of a handclap—could cause the boundary layer to separate long before it reached the peak of its crucial ascent. Like a surgeon making discreet incisions, Mark continually probed the brains of his computer until the airfoil's surface was just imperfect enough to fly.

Late one night Juan found his way over to Building 37 and wandered into Mark's office. With endless design choices for their wings and bones, the two engineers scrawled all over the chalkboard. Mark had decided to make different airfoils for three distinct sections of the wings. The design called for wings stretching over one hundred feet, and because the length forced him to account for the variation in Reynolds Numbers from the fuselage (600,000) to the wing tips (200,000), every point along the leading edge had to be tapered and customized. Juan had chosen a spar design shaped like a triangle, rather than a circle, and he'd even built a test section, which he showed Mark.

"It was a bitch to build and it looks like shit," Juan said. "The problem is, I know why it looks like shit, but I can't quit making it look like shit." It had been a good idea. Theoretically, the triangular spar with rounded corners would be strong enough to delay buckling, and the three sides formed the proper geometric area to handle twisting and bending forces. On paper, it was fine. In the lab, only one word could describe it—Woof.

They needed a new idea. A design to save weight but that could still handle the bending, twisting, and shearing forces from Mark's wings. The bones had to be pliable. They needed to bend like a graphite fishing rod during flight, but they had to be stiff enough not to break in midair. They had to be light enough so the pilot could power the airplane, yet strong enough so the ground crew could touch the parts without crushing them.

"Wait a minute," Juan said. "Let's do this." He drew a big circle on the chalkboard and then two little circles, one on the top and one on the bottom. It looked like a Christmas tree ornament. It was the stupidest design Mark had ever seen. He laughed and then Juan laughed. When they quit laughing they left the office and took an elevator down to the basement where, for the next two days, they worked on a test section for a spar made of one large carbon tube and two small cap tubes, one on the top and one on bottom. When they finished gluing it together, the piece looked like a Christmas tree ornament.

Faculty members who saw the spar were appalled. The little cap tubes, on top and bottom, had been made of carbon sheets laid out at zero degrees (wrapped so the fibers were parallel to the shaft or mandrel; see photo in first photo section). At such an angle the wing spars would provide maximum bending strength—they'd allow the wings plenty of flexibility to curve up without breaking. The large center tube, whose walls had been made with alternating plies wrapped at opposing angles of +40 degrees and −40 degrees, served as the torsion tube. It would hold the wing straight and true without twisting or giving in to the natural bending moments created by the airfoil. It sounds good, faculty members would tell Juan, but there's just one problem. Two circles meet only at one point. There's no flat surface to hold the pieces together. How are you going to glue circular tubes?

"Don't worry," Juan would say. "It works."

The faculty hated this idea of three tubes kissing. It was crazy, it was laughable.

Juan and Mark went right to work, though. The faculty's skepticism was an inspiration, the subtle taunts only another challenge to overcome. No one outside the working group could lay claim to their designs or deny the independence of their spirit. This airplane, they'd say confidently, is going to fly.

10

He was burrowed in an office choked with paper. Icons from the Far East and ancient Greek cultures dressed the walls; glass-encased bookshelves groaned under the weight of frighteningly academic texts about exercise physiology. The room might have been the nest of a squirrely academic. But Ethan Nadel's workplace served as much as an athlete's retreat as an intellectual's sanctuary. In this room Ethan had built an impressive redoubt against the encroachment of middle age.

At midday the senior research fellow at the John B. Pierce Foundation often could be seen standing shirtless by the window, wearing only shorts and tennis shoes, dripping sweat and panting after a two-hour jog through New Haven. Physiological readings and intuition would tell him what he needed to know. At forty-seven, Ethan had qualified for the Boston Marathon every year for the past six years and every year his time had improved; 2:56 was not bad for a guy who still liked his beer and potato chips. "Two fifty-six is really respectable," Ethan said. "It even surprises me. I'm not built for long-distance running."

Because of his rugged stature, a rumor once went through the working group that he'd posed for a statue of Neanderthal Man on the Yale campus. He would have made a fine model: he was short and stocky, with a heavy, bristly beard, a strong jaw, and dark, hawklike features. When he ran, his thick body waggled side to side so awkwardly that those who joined him sometimes found it remarkable he could endure the jostle of his own stride. But with the benefit of a strong will and an insider's knowledge of endurance exercise, Ethan outpaced most of his companions. In 1983, when he went to the twentieth reunion of his graduating class at Williams College and entered a road race mixed with alumni and undergraduates, Ethan finished fourth. "Use it or lose it" was his attitude, a homily based as much on science as personal experience. Ethan was his own guinea pig.

By the time he joined the working group at MIT in the fall of 1985, the physiologist's name was well known in academic circles, yet Ethan had eased into a contented, contemplative routine. If their paths had crossed five years earlier perhaps, it was unlikely that Ethan would have given up 20 percent of his time to tinker with an HPA's engine. A few years earlier, he'd wanted to be the best in his field and the challenges of physiology filled his life. But increasingly, as he grew older and learned about extending the boundaries of age, he enjoyed life outside the lab.

In 1985, he might not have been the very best in his field, but he was certainly among the best. A team of talented research assistants worked under his direction. Grants came easily to his office, and demand for his skills and insights sent him traveling across the world. He lived comfortably, too, in a three-story, one-hundred-year-old farmhouse in the pretty town of Guilford. Ethan had a secluded acre of land to enjoy, a pond, a sailboat, a vegetable garden, a pot-bellied stove and a cord of wood, a vintage Mercedes, and a couple of Volvos to toy with out in the garage. He had a mechanic in New Haven who did good, inexpensive work in the shade of an oak tree, and an attractive girlfriend who sang in a Westport church choir. Perched high on the plump seat of his yellow Mercedes, he could be seen every afternoon driving home, coursing the streets leading from his lab to his house near the Connecticut shore. Ethan's head would shift side to side as he caught glimpses of the world outside slowly slipping by. He certainly hadn't discovered the fountain of youth, though it seemed as if he had found the next best things.

By middle age, Ethan could carry on an incisive conversation with a professor of literature about Herman Melville's intuitive ability as a biologist. He visited European art galleries, studied the architecture of Gothic cathedrals, and followed the development of stone circles to the Hebrides. The enthusiastic group of young engineers who interrupted his life with an immoderate claim that a human could fly from Crete to the mainland of Greece—under his own power—caught Ethan at a peculiar moment of personal, if not scientific, discovery. He found himself unexpectedly intrigued and excited.

* * *

The engineers needed a mechanic badly. Parky's earliest airplane designs included an HPA that dawdled along over the Aegean for up to eleven hours on its sixty-eight-mile course. (Flying slowly offered benefits to the human engine: power requirements and speed worked against each other. In Parky's program, every time they increased the speed—even incrementally—they increased the engine's effort by an enormous amount.) They could accurately simulate the weights of carbon and Mylar, the lift on the wings, length of the prop, and even estimate the number of external wires the plane needed. But nowhere in Parky's design program did it account for engine failure. The heaviest and most complex piece of their airplane was still a mystery. At two thirds the weight of the whole package, the engine comprised the part they called "the black box." Parky's data, extrapolated from endurance tests up to two hours long, showed that their human engine would still be pedaling as heartily after eleven hours as at four. They didn't know a thing about predicting glycogen burnout.

Unlike most airplane engines, next year's HPA power plant wouldn't be much better than what was already on the shelf. The technology of the human being remained pretty constant. But there were certainly factors that limited the body's performance, and some types of bodies seemed to suit an HPA better than others. When the engineers asked Ethan how much power they could expect from the black box, he asked in return, "How much power do you need?" Nobody knew, quite.

Ethan realized he'd have to quit thinking like a Yale physiologist and think more like one of the boys. If he could make the transition from science to engineering, he'd find solutions. The guys at MIT modified a stationary recumbent bicycle called an ergometer in Ethan's climate-controlled lab and crowded the room with blood and lung monitors, EKG machines and oxygen tanks. There'd be no controlled experiments, no esoteric questions about why the human engine functioned as it did. The idea was to take a very inefficient machine—the human body—and make it work as well as possible. Ethan's lab became a workshop for testing the best engines he could find and offering weights versus power measurements to the engineers. He began to think of the human body as a car engine, comparing the combustion and cooling processes, fuel and oxygen requirements. How much power did they need? Oh, he'd say, about one-third horsepower.

Just like any other part of the airplane, the weight and efficiency of the engine had considerable effect on the power requirements and design of the craft. The world's strongest weight lifter, for instance, wouldn't suit them because a weight lifter's power wasn't aerobic. Presumably, the lifter's size would require heavier carbon tubes to withstand the additional mass, and the combined bulk would jack power levels beyond human endurance. Ethan knew the engine, like the spars, had to be extremely efficient—the person had to be both lightweight and powerful. Because the engine also would have to carry his or her own weight during the pedal-powered flight, Ethan looked for what the engineers called "specific power," expressed best in a power-to-weight ratio. For instance, using Parky's sizing optimization program, the engineers estimated that with a 150-pound engine and a preliminary airplane weight of 70 pounds, they'd need approximately 200 to 250 watts of continuous power to keep the wings airborne. Converting power requirements to watts per kilogram of body weight, Ethan assumed he needed to find an engine—a human being—who could produce between 3.0 and 3.5 watts per kilogram. (The math is simple. Convert pounds to kilograms: 150 pounds = 68.18 kilograms. Then derive the ratio of watts to kilograms: 200 watts/68.18 kilograms = 3 watts per kilogram.)

Was it feasible? A survey of physiological literature over the last fifty years provided few answers. Besides offering almost no data on human endurance beyond two hours, many times Ethan questioned the scientific validity of the studies, as the results rarely accounted for such factors as body weight, level of training, and environmental conditions in which the tests were performed. In a report to the working group, he explained, "Although the historical record is useful for establishing rough bounds to the problem, it is of little help in establishing the engineering feasibility of the flight. For that, we must turn to first principles."

The human body, like a car's engine, combines fuel and oxygen to produce energy and waste byproducts. An ordinary adult socks away an enormous amount of fuel at various places in the body. For example, most of this potential energy is kept in adipocytes, or fat storage cells. An average-sized person has enough potential energy saved in adipose tissue to run a hundred miles. The rest of the energy comes in the form of glycogen, stored in the liver and skeletal muscle. Unfortunately, this fuel depends on nervous and endocrine reflexes to make it available for combustion with oxygen. Worse, the body's supply of oxygen is relatively small—about a liter, only enough to sustain moderate exercise for thirty seconds at best. To make the most of the body's fuel, the human engine needs a continuous and abundant flow of oxygen transported from the lungs to the heart to the muscles. Even then only about 24 percent of the stored fuels can actually be converted into mechanical work. The remainder of the fuel—76 percent—converts to heat during the metabolic transfer.

To help the team imagine the physiological challenge better, Ethan drew a little diagram that he passed out at group meetings, which looked like this:

$$\text{CARBOHYDRATE} + \text{FAT} + O_2 \rightarrow \text{ENERGY} + CO_2 + \text{HEAT} + \text{WATER}$$

From out of the equation, Ethan identified the limiting factors for their pilot. Most importantly, they needed a person who could deliver oxygen efficiently and in large quantities. Since most of the energy would be converted to heat, in order to get 3.5 watts per kilogram out of the engine, the person would have to maintain a fuel oxydation rate of 14.6 watts/kilogram. That meant whoever sat in the pilot's seat would have to consume continuously, for up to four hours, 44 milliliters of oxygen per minute per kilogram of body weight. That was a lot of air.

For the average man, the maximum uptake of oxygen was approximately 50 milliliters. That is, for most men 50 milliliters of oxygen would be all their lungs could absorb per minute. An ordinary man pedaling at his maximum aerobic capacity might be able to keep the plane running for a few minutes, but there'd be no way he could fly it for a few hours. Put an average human engine in the fuselage and, by most estimates, the legs would quit cranking in a matter of minutes. Ethan knew that when an ordinary person pushes his or her energy production level above 60 percent of capacity, the body relies increasingly on what are called anaerobic processes. As it reaches its maximum energy level, the body's metabolism changes and its waste products—lactic acid and hydrogen ions—cramp the muscles and make them feel like they're burning. The metabolic costs, at such moments, couldn't be overcome in the air, where the pilot couldn't stop pedaling or even slow down to rest. Ethan's calculations showed the engine requirements would be nearly the same as those required to pedal a bicycle over level ground at a world-class speed of 22.2 miles per hour. Clearly, not just any human would do. They needed an athlete in the fuselage, someone whose body already had adapted to a high level of aerobic training.

Ethan found a partner at MIT to help in the search for a perfect engine. A young, untenured assistant professor in the aeronautics department at MIT had come to most of the working group meetings and offered to work part time on the project. Steve Bussolari was a "human factors" expert at the 'Tute, an intense scholar whose academic terrain encompassed pilots' and astronauts' physiological responses to supersonic flights and unusual atmospheric conditions. Bussolari might have been a reflection of a younger Ethan—Ethan during his ambitious, strictly academic phase—and one could sometimes hear Ethan talking privately over the phone from Yale to Bussolari at MIT offering some advice about the politics of the academy. But just as much, the two seemed to share some undefinable but characteristically independent outlook on life. When they met, on occasion, they would go for a run and share a couple of bowls of bananas and bran. Sometimes, they'd draw a few beers together, too. They maintained a professional profile among the Daedalus team, but one always suspected they were having as much fun as Juan and Mark. Ethan and Bussolari became a pair, another team inside the team.

During October 1985, male and female athletes traipsed into Ethan's lab on weekends. He and Bussolari would test triathletes, field hockey players, national-class wrestlers, and former cyclists from the U.S. National Team. Within ten or fifteen minutes, Ethan could tell if the candidate would make a decent engine. Bussolari would sit him down on the ergometer, clip his nostrils with a pin, stick an oxygen hose in his mouth, and make him ride to the cadence of a metronome set on allegro while Ethan increased the loads on the pedals. Once the person hyperventilated "significantly" (a respiratory exchange of carbon dioxide to oxygen greater than 1.15), the test ended. Their so-called VO_2 Max test was a unique torture, consistent with an engineer's favorite structures experiment, "test to failure." But instead of hearing a crack or explosion, the two professors watched their engines turn red and then blanch. Ethan listened behind a bank of instrument panels as they heaved, choked, and spit during the final seconds of the test. Bussolari would stand at the ergometer and yell, "Push! Push! Push! Push!" to drive them to failure, and then Ethan would disappear into an office to read the outputs from the heart-rate, power, and oxygen monitors. What they wanted were athletes who, when they reached maximum oxygen uptake, produced power at over 4.29 watts per kilogram. At 70 percent of that limit, Ethan felt confident that his human engine could fly comfortably for several hours without cramping, overheating, or running out of fuel.

With the engine tests under way, Ethan and Bussolari discussed the potential problem of the pilot's overheating, another limiting factor in the equation. If they expected their engine to produce 200 to 250 watts of propulsive force, they had to design for overheating. At the expected power, even under moderately cool conditions, the body would produce 600 to 1,000 watts of heat, enough to raise the core body temperature 1 degree Celsius every five to eight minutes. "Imagine sitting in this fairly small cockpit with ten hundred-watt bulbs in there," Ethan explained. "You have to have air flowing through the cockpit. We can sit down and figure out how to do it without taking too much of a drag penalty on the airplane, but you've got to provide air vents. If the engine isn't sound, it doesn't matter how aerodynamically good the airplane is."

Ethan volunteered to do thermal tests, heat-sensitive video tapes of athletes pedaling the ergometer, that would show which part of their bodies most required cooling. The results could be used in pinpointing precisely where vents should be cut in the clear, plastic fuselage.

The two professors also identified another important variable having to do with overheating: the weather. Ethan would worry about how to cool the engine with water, and Bussolari decided he'd undertake a study of the Greek climate. Temperatures that exceeded 64 degrees Fahrenheit, humidity levels above 70 percent, and insolation of the cockpit (solar heating) could stress the pilot beyond endurance. Not only would the team have to search meteorological records to see which weeks of the year were the least windy, but they would have to identify dry days when the thermometer stayed below 64 degrees during the early morning hours.

With the help of Bussolari, Ethan himself became a pretty good mechanic. The weather window narrowed. A search began for fuselage skin that would reflect the sun's rays rather than absorb its heat. Never before had so much effort gone into the care and handling of the human airplane engine. Now all they needed were more specimens and more tests.

Only two athletes qualified. Rudy Sroka, a two-time member of the U.S. National Cycling Team, gave a burst of power during his VO_2 Max test in Ethan's lab and exceeded the engineers' expectations for an engine. At the end of his ten-minute trial in mid-October, the data showed he'd produced 324 watts. Accounting for his body weight (161 pounds or 73 kg), his 70 percent power level topped the baseline requirement at 3.11 watts per kilogram. Every serious cyclist in New England knew Rudy. He'd competed in the Olympic Trials in 1976, and in 1980 he went again to the U.S. Olympic Training Center to ride. Although he'd slacked off the competitive pace in recent years and made his living selling bikes for an Italian importer rather than on the racing circuit, he remained a powerful regional cyclist and triathlete. After his test in October, Ethan asked Rudy back to the lab in January for a four-hour endurance ride on the ergometer.

The other candidate was an unknown. At twenty-nine, Lois McCallin spent most of her time working as a short-term municipal securities trader and a software designer for Fidelity Investments in Boston. Since graduating from Drew University in New Jersey, she'd spent eight to ten hours a day sitting in front of a computer screen. Only in the last two years had she vowed to break the routine and test herself physically. In 1983, Lois made a New Year's resolution to start exercising, and slowly she'd developed a regimen that included running, swimming, cycling, and flying airplanes. "I just decided that people should do things with their lives, and I hadn't been doing things," she said. "I had been working my nine-to-five job, but I had real ambitions."

She'd started conditioning by commuting to work eighteen miles every day on her bike. She ran her first marathon in 1983 and by the next year she was training for triathalon events, combining swimming, running, and cycling. She was an extraordinarily slow swimmer, but set a fast pace on a bicycle. By the fall of 1985, she usually placed in the top ten at regional triathalons. Although she still referred to herself as a "neighborhood-class" athlete and never dreamed of competing against real athletes like Rudy Sroka, she was excited by her modest successes. When she learned about

Ethan's tests for a human-powered airplane engine through an article in *The Boston Globe,* Lois clipped the piece out of the newspaper and stuck it to the corkboard in her kitchen. She looked at it every day until she mustered the nerve to call for an appointment.

"I've never been tested in a lab," she told Bussolari when she called to volunteer for a VO_2 Max test. "I've never had a concrete measure of my physical abilities."

Lois's VO_2 Max score convinced Ethan that his ideas about specific power had been correct. At 235 watts, Lois didn't produce the most power among the candidates nor did she score particularly high in terms of her maximum oxygen uptake. But the relative value of her power—the ratio of power to weight—nearly equaled Rudy's. Weighing one hundred twenty-two pounds, Lois McCallin proved to be a formidable power plant. When Ethan invited her to try a four-hour test in January, Lois bought a stationary bike and set up a new training program for endurance cycling in her apartment. She rode several hours a day. "I was motivated," she said. "I figured everybody wanted to be the pilot for that airplane."

Lois drove down to New Haven and met her brother one Saturday in January 1986 carrying a boom box and a load of cassette tapes to get her through the four-hour test. They walked up the steps of the Pierce Foundation, then up to Ethan's lab on the second floor. Rudy's girlfriend had been sitting outside in a waiting room for two hours while Ethan and Bussolari went in and out of the lab, checking chart recorders that spewed out graphs showing Rudy's heart rate and oxygen levels. Lois went into the lab to watch, but she didn't stay long.

Rudy looked pretty worn, stripped to his shorts, grinding the pedals on a tough race to nowhere. He had EKG electrodes stamped to his bare chest, an oxygen hose stuck in his mouth, a catheter plugged into his left arm. The catheter was held in place with surgical tape, and Ethan took blood samples by pinching a little stopcock in the plastic tube that led to the vein in Rudy's arm. A fan blew across his upper body to cool him and the metronome clicked a quick cadence, which Rudy struggled to match at 76 rpms on the ergometer. Ethan and Steve checked the monitors continuously, drew blood samples, and encouraged Rudy to keep pedaling. His heart rate was rising, though, and not long after Lois saw him, he started to fall slightly off cadence. After three and a half hours, Rudy's heart rate climbed to 180 beats a minute and his legs cramped. He didn't even think about stopping. His body just gave up.

Shaken by the cyclist's failure, Lois called her brother into the lab to stay with her for the four-hour test. She'd trained as long as three hours on her stationary bike at home, but never pushed to four. Ethan and Bussolari hooked her up to their machines and let her sit for half an hour on the ergometer before she started pedaling. They took heart-rate, blood, and oxygen measurements at her resting position. And then the metronome began to play.

As Ethan slowly tightened the belt around the wheel of the ergometer, taking it up to the 3 watts per kilogram power level, the pedals became increasingly hard for Lois to crank. After an hour on the cycle, she ate a banana. Her brother slapped a Dire Straits tape into the boom box. She drank a bottle of water. Then another. She ate an apple. Her brother switched tapes. Lois kept cranking.

After three hours and forty-five minutes, Ethan and Bussolari examined the charts coughing from the recorders and saw that Lois's heart rate had never gone above 160 beats per minute. She and her brother filled the lab with the sounds of rock-'n'-roll and a swishing cadence that never faltered. "Play Springsteen," she said. "No, wait. Play Meatloaf—'Bat Out of Hell.' "

After three hours and fifty-five minutes, the two professors ran into the lab with a six-pack of beer and started to crack the tabs.

"Don't . . . not yet. . . ." she gasped. "Wait until I'm finished. I want one of those."

Lois and her brother spent the night in New Haven. The next morning, though her legs ached, she joined Ethan for a fourteen-mile run. In his findings to the working group, Ethan reported on the outcome of his tests:

> The second test subject (a female triathlete) was able to complete the four hours easily and could, by her own estimate, have continued for another thirty to sixty minutes. We conclude that a well-trained endurance athlete can exercise at seventy percent VO_2 Max for four hours in optimal conditions. It appears from this preliminary study that fluid replacement will be an extremely important factor to ensure

the maintenance of the physiological steady state. Further, given optimal conditions, neither fuel transport nor oxygen transport to muscle should directly limit performance in this range.

Ethan had just found himself an engine.

11

One by one, the wild ideas vanished. The eight-hour pedalfest to Ikaria. The two-tailed Icarus flight. The 8,000-foot cliff launch. The triangular spar. On a 2,400-mile road trip back and forth from Burbank to Amarillo to pick up a sailplane, Parky and Gup finally had a chance to argue one another past the point of make-believe. No longer interrupted by John's schedules or working group dynamics, they hammered each other's whims into oblivion. After three days together on the road, they trashed the pusher prop design, the speedplane idea, the notion of building a hinged propeller, and the fuselage modeled after a human-powered land speed vehicle called Vector. "We have to consider all the weird ideas before we settle for the conventional ones," Parky explained to Gup.

Well, of course. And so it was with every other detail, whether it came from Juan or Mark or Ethan or John. Skeptics demanded proof, and those who were called on to defend their ideas had to quantify. For eleven months, the whole team engaged in the same detailed, scientific sort of risk assessment. Even their classical antecedent became clear; no one referred to the team as the Icarus Project anymore. In fact, by the end of January 1986, there was only one problem standing between the working group and the workshop where they'd build the Daedalus prototype—an empty ledger sheet.

The final stress test took place in the Langford household. John and Barbara had a six-month-old son, Ellis, to tend; John had a dissertation to complete, and a part-time job at the Institute for Defense Analyses studying a hypersonic airplane. The Daedalus Project was beginning to inflict itself on the family order. Barbara, who'd not only helped fund the earlier HPA projects but worked at times to support them while John was in school, assumed the rules would be different for a $690,000 airplane. John agreed. First, months earlier, he'd set a ground rule: He would not raise money. Dealmaking duties fell to the finance committee, to guys with credentials and connections—MIT's associate dean of engineering, Jack Kerrebrock; former NASA administrator Bob Seamans; National Air and Space Museum administrator Brian Duff; and aeronautical entrepreneur Jim Draper. Second, John promised Barbara the family would not make financial sacrifices for the project. They wouldn't bankroll it, and John wouldn't let Daedalus prevent him from taking a doctoral degree in the spring. Fine, Barbara thought. If they can't raise the money, life goes on.

For months members of the finance committee engineered deals. Letters went out to a hundred major companies and foundations across the country. Kerrebrock, Seamans, Duff, and Draper made calls to personal friends in the top echelons of IBM, Boeing, McDonnell Douglas, Raytheon, and Boston University. The old boy network, John told Barbara, is much more pervasive than people would ever believe. But even it had limits. A personal call won them a hearing, but no guarantees. Late in the fall of 1985, Draper dropped out when his own aerodynamics research company needed more of his attention. Kerrebrock and Seaman's executive friends shared their enthusiasm for the Daedalus flight, but they had to answer to marketing departments and stockholders. The second phase of the project—the building and testing of a prototype airplane—would cost almost $300,000, and no one wanted to take a chance.

During the same week Ethan proved in his lab that a human could make the four-hour flight, John met Paul MacCready for lunch in Palo Alto after a conference on high-altitude endurance airplanes. John hesitated to ask for advice, but between drinks, the man who still had the reputation as the world's leading innovator in human-powered flight offered sobering news. His words stung like an oracle's warning.

"Our numbers predicted that the Albatross could have gone a heck of a long way if there had been good weather," John told him. MacCready, bespectacled and reserved, ventured no comment. "It could have flown thirty-five miles on a cross-Channel flight, if you discount the head wind factor. The actual performance we're looking for, in the Daedalus recreation, is only a factor of two times the Albatross flight."

"The airplane will be easy," MacCready said. "Getting there will be a nightmare."

Sure, John thought. I've managed the logistics. He had photographs of possible launch and landing sites. He'd engaged the efforts of a Yale University classicist. A dozen undergraduates at MIT had signed up to build test components for the prototype airplane. Mark Drela and Debbie Douglas had traveled to England to meet with Clive Hart, an international figure in the literature and history of flight. He'd made preliminary arrangements for a film of the flight with a major documentary producer in Boston. He'd even sent Steve Bussolari on a trip to Greece to meet with meteorologists, to survey flight routes, and make connections inside the Greek ministries of national defense, transportation, press and information, culture and civilization. As a manager, he'd steered the project to satisfy their benefactors at MIT and the Smithsonian. They'd been waiting patiently for more than a year to see the project meet its stated goals: to make the Daedalus Project a cross-cultural, interdisciplinary exchange; to conduct scientific research in aeronautics, physiology, and meteorology; to share their efforts with students and the public. He'd done his job. His engineers had done theirs. The only slack fell, he thought, where he had no control or authority.

John considered MacCready's warnings, as well as some advice about relying on a professional to raise money rather than depending on the goodwill of the old boy network. On the first Sunday in February, he and Barbara discussed their options. That afternoon, he planned the final meeting of the steering committee, and mailed a memo that described their dilemma. Within two days the following note appeared on the desks of steering committee members, Walter Boyne, director of the National Air and Space Museum; Eugene Covert, chairman of the MIT Department of Aeronautics and Astronautics; James Draper, president of Aerodyne Air and Research, Inc.; Jack Kerrebrock, associate dean for the school of engineering at MIT; Robert Seamans, senior lecturer at MIT's aero department; and Brian Duff, former director of public relations for NASA, now associate director of the Air and Space Museum in Washington:

This memorandum is to confirm the meeting scheduled for Friday, February 7, 1986. The primary task of this meeting will be to discuss what plans, if any, seem desirable for continuing the project after the release of the Phase 1 report next month. We were turned down most recently by the Honda Foundation (we were "too entrepreneurial," they preferred to give to "charities"), so we have no funding commitments beyond the Phase 1 budget.

Unless someone pulls a financial rabbit out of his hat before this meeting, I believe we have no choice but to abandon any plans to begin Phase 2 during 1986. This is especially unfortunate in light of the continuing good results from our research program. On January 25 one of the athletes in the test program completed a full duration flight-power ergometer run in the lab at Yale. The first proof test of the spar test section ended with a shear failure at 3.5 times flight loading. Fully detailed weather data has arrived from Greece. The technical evidence all points to a solid conclusion of "feasible."

All this is moot without the financial resources to proceed. The question we must decide on Friday is how the project should proceed once the final report is completed. I believe our options include these:

1. *The project, while technically feasible, is too expensive to proceed with. We should effectively end the project with the release of the Phase 1 research report.*
2. *The current fundraising approach is sound and should be continued for another 6 months or so.*
3. *It is time to hire a professional fundraiser.*
4. *A formal proposal should be submitted to the Greek government asking them to fund the project.*

Personally, I believe that option 1 is not justified. Each of the 6 U.S. America's Cup syndicates has raised $7–$13 million from some of the same companies we have approached. Five of these entries are sure losers, especially compared to the reception of Daedalus.

I think the second option, continuing our present course, should now be ruled out. We have tried all the obvious targets and it can only get more difficult from here; few members of the Committee (understandably) seem very enthusiastic about the prospect of continuing to carry the full burden of the fundraising. Choosing this option would mean that we would depend heavily on the Air & Space Museum. As you know, the Museum is faced with severe pressure on its federally-supplied funding, which may in effect increase the competition for Daedalus and may well lower its priority in the Museum.

The third option was the one taken by MacCready on Albatross and is the one option I now frankly prefer, although I don't know much about how it would actually work. As I understand it, we would hire a professional fundraiser, cover their direct expenses, and give them a percentage of what was actually raised. I doubt we have sufficient funds in the Phase 1 budget to cover this.

The final option makes sense, but would surely cost us some flexibility and control. We would be constrained to using a Greek pilot and perhaps lose final authority over such things as when we fly.

I hope that each of you will consider the entire situation carefully and come prepared to make some definite decisions at the meeting. Since this weekend will probably be the last working group meeting for a long time, it is essential that definitive guidance be provided from the Steering Committee.

When Walter Boyne read the note, he quickly called Brian Duff into his office. They discussed the merits of the project privately. At the time, the museum had suffered considerable internal strife over its involvement in fanciful projects—providing assistance to Dick and Burt Rutan's round-the-world Voyager flight and MacCready's mechanical Pterodactyl—and some of the historians felt these had been, frankly, commercial diversions from the scholarly nature of the museum. There wasn't a lot of discussion; Boyne's directive was clear. On Friday afternoon Duff left the Smithsonian to catch a flight to Boston, carrying a heavy set of marching orders from his boss: Disengage from the Daedalus Project, immediately.

The group that gathered in Kerrebrock's office at four-thirty Friday afternoon couldn't have been more ambivalent about their choices. John wanted out, at least for a year, so he could earn his Ph.D. and keep his promises to Barbara. Kerrebrock and Seamans wouldn't accept John's proposal to hire a professional fundraising team. Draper reluctantly explained he could no longer offer much help. As for Duff, he listened quietly and didn't speak.

As John summarized the results of the team's eleven-month research effort, an image kept slipping in and out of Brian Duff's mind. Years ago, he recalled, there was a cartoon character from the old *Saturday Evening Post,* a little guy who worked for the Earthworm Tractor Co. A bumpkin named Botts. Alexander Botts. Botts the troublemaker, Botts the traveling salesman. He recalled the time when Botts cut red tape for his customers, and a great cry rang from the home office: "Botts, you idiot, don't you know you've just violated every company rule!" And Botts replied: "No, I haven't. I'm only doing what the president would do if he understood the situation."

When John finished, he reiterated his support for hiring a professional fundraiser, and putting the project on hold for that academic year. He could use the time for his dissertation anyway. "No bucks, no Buck Rogers," John said.

Suddenly, Duff spoke up.

"It's too early to quit!"

Everyone was startled.

"I think we've got a good project going here," Duff said. "We haven't reached a point where we can make a rational decision to stop. We're just a little discouraged. And we've got ninety days left. These guys haven't even written up the results of this Phase One study. I mean, MIT and the Smithsonian have paid for that study and we don't even have it yet. Forget enthusiasm. We can justify this project on the basis of our original deal, and it'll give us one more chance to find a sponsor. I'll tell you, we've got everything to gain and nothing to lose. This is a great project from beginning to end, and I wouldn't mind having two or three more like it before I retire."

At dinner that night at the MIT Faculty Club, Kerrebrock told Mark, Juan, Gup, Ethan, and the rest of the working group about the steering committee's decision. John smiled. He'd accepted the outcome, and he'd even acted grateful. But he felt compromised, and he was privately pretty angry. Sure, that's easy for them to say, he thought. Push ahead? Bullshit! They don't have my family commitments. They're not facing a dissertation committee. Who the hell's gonna raise money now? This is just going to muck up my life.

That night John revised the outline for the working group's final report. Its collection of papers—hitherto, scientific documents on meteorological conditions in the Aegean, an exegesis of Daedalus mythology, an aerodynamics analysis of their designs, a discussion of Ethan's physiological tests—wouldn't wind up in the Smithsonian as an archival document, as originally intended. Instead, John would mold it into a sales tool, a snazzy series of short technical papers and a primary report that would "position" the project for marketing directors at specific Fortune 500 companies. John tossed out his ground rule not to be a salesman, and he took over the project's finances. He had less than three months to raise $300,000.

MIT's own ground rules couldn't be so easily altered. In the partnership agreement between the 'Tute and the working group, Jack Kerrebrock explicitly insisted that the Daedalus Project would not be allowed to compete with normal fundraising activities of the university. When he realized how serious John had become about raising the money himself, Kerrebrock took him aside and explained the limitations.

"Let me go back and make this point, John," he said. "It's very important: We're not going to allow the project to siphon off money that would go for chairs, scholarships, general research, or anything else for which there's a very well established system of solicitation. That's the bargain we've made.

"Frankly, the reason for that is that the cost of the project is high in comparison to its perceived research and educational benefits. Now, what the project gets in return is valuable. What the Institute puts into it is its reputation. And I'm responsible for signing off on that. So when you approach these potential sponsors, it has to be from the basis that the project has an image-building value. Advertising, to put it crassly, because of its media appeal and charm. Don't be surprised if you get turned down. They'll say it's charming and wonderful, but it'll be hard to get a dollar beyond that. Everybody agrees about the charm. The thing you've got to do is convince them that it'll bring good publicity."

"It's a choice between buying a thirty-second spot during the Super Bowl or supporting us," John said.

"You've got it."

"Okay. Who's left on the list?"

They already knew that the obvious sponsors would brush them off.

When Seamans and Draper first approached the chief scientist at IBM, the company had just announced it would sponsor a television series on Leonardo da Vinci, and it seemed logical to propose Daedalus as a companion piece. The response from public relations, though, was, essentially, "We're a computer company. Get an airplane company to sponsor it." When Kerrebrock floated the idea with executives at Boeing, they loved the project. But unless they could contribute from their educational fund, the flight from Crete couldn't be justified to shareholders.

"They said, 'Okay, it has to be done from the public relations side,' " Kerrebrock explained. "As it turned out, they figured no one needed to be convinced that Boeing knows how to build airplanes."

The Honda Foundation had been their biggest disappointment. The guidelines for its grants said it supported university-based scientific projects characterized by a youthful spirit, imagination, creativity, humanistic values, and forward-thinking approaches. Juan, John, Kerrebrock, Duff, Steve Bussolari, and Professor Covert all made a strong pitch to the foundation's representatives in December 1985 at a meeting in Cambridge. "According to their brochure, we were perfect for it," John said, "but that wasn't right. It turned out they really wanted to give their money to charities. . . . We were competing against people with artificial limbs in Indiana and against a center in New York that helps the children of heroin addicts. One of my personal ground rules is never to compete against charities, and we never would have applied if I'd known that. But I was still really ticked off because we put a lot of energy into making a good proposal, and we were counting on that money."

Only two companies from the original list remained untapped. Union Carbide Corporation, in Danbury, Connecticut, which produced the carbon composites Juan would use to build his spars, was a natural. E.I. DuPont in Wilmington, Delaware, produced other materials that would go into the Daedalus airplanes—Mylar and Kevlar, in particular. Since DuPont also had sponsored MacCready's Albatross flight, John expected the company might have an interest in supporting Daedalus, too.

Then there was the beer company—Anheuser-Busch Companies in St. Louis. Busch had been Seamans's idea. Just before Thanksgiving Seamans had called Sandy McDonnell of McDonnell Douglas, and McDonnell, like every other aeronautical engineer who'd heard of the project, loved it. John suspected the reason Sandy McDonnell didn't agree to back them, though, was because it was the time of the $700 screwdriver. The company had taken a pounding on Capitol Hill for the perception that the aerospace industry had been spending taxpayers' money foolishly, and it wasn't the right moment to sponsor a crazy scheme to fly from Crete with seventy pounds of Mylar, Styrofoam, and graphite tubes. After Thanksgiving, though, McDonnell returned the call to Seamans with a tip.

"I don't think it's appropriate for us to fund it," McDonnell said. "But I had dinner last night with a guy at Anheuser-Busch, and they might be interested."

August Busch, chairman of the board of the nation's largest brewery, ran a company best known for its Clydesdales. But he really enjoyed hunting, skiing, driving racing cars, and flying airplanes. In fact, Busch was privately known among his closest colleagues as probably the greatest flying enthusiast, outside of an aerospace company, in the corporate world. And, yes, Busch was interested.

With April 9, 1986, scheduled as the release date for the Phase 1 report at the National Air and Space Museum, John rushed to set up meetings in St. Louis, Danbury, and Wilmington. Juan and Brian Duff would go with him to visit Union Carbide. Juan agreed to make the pitch with him at DuPont's corporate headquarters. Ethan and Duff said they'd meet him in St. Louis. "Don't forget," John told them. "No bucks, no Buck Rogers."

He'd thought Ethan would turn the trick, but Duff made the most difference during their dog-and-pony shows. Silver-haired, lean as a bullet, handsome as Gary Cooper, Duff lent them an image of class, status, and clout. Maybe it was his twenty years with NASA, his international tour of duty with the astronauts, or his contacts on Capitol Hill made as a journalist during the Kennedy years. When he spoke, John said, he sounded like Arlo Guthrie's Sheriff Obie talking to the blind Judge in Alice's Restaurant—Duff didn't care if no one wanted to look at the twenty-seven eight by ten color glossy photographs, he'd just keep talking.

The first pitch Duff made was at Union Carbide. He talked about NASA, he talked about Daedalus, he talked about the future and the dream. "When he was finished," John told Barbara later, "everyone was ready to stand up and salute the flag. But they couldn't do anything. They said, we love this, but we're on the block. Our division's being sold to Amoco, and that deal's gonna take another six months to settle. They said they'd be real interested in proposing the idea to the Amoco management if they're still running the composite operation then. That's not much good to us now."

At DuPont, Juan and John met with the man who'd struck deals with MacCready. Dick Woodward, in the external affairs department, led them into an office filled with relics of the Gossamer Albatross. The Albatross project had been Woodward's greatest achievement at DuPont. On a $180,000 investment, the company estimated that the Albatross flight had brought worldwide exposure for DuPont, the equivalent of more than $300 million in advertising. Yet Woodward said he wasn't interested in taking the same risk again.

"Why should I compete with the high point of my career?" he said. "I've had two heart attacks and I've got nothing to gain from another project like this. If you fail, I lose. If you make it, at best I break even."

At two o'clock on March 13, John, Ethan, and Brian Duff drove past the Clydesdale stables, the yeast plant, the brew house, and entered the Anheuser-Busch corporate headquarters in St. Louis. "It doesn't do you any harm to have a guy with a little gray hair on your side," Duff whispered to John.

As they got off the elevator and walked through the offices, they met a series of men who looked like former linemen off Ohio State's football team—thick-bodied midwesterners with straight backs and big teeth. "I always wondered what happened to those guys," John said to Duff, who was thinking the same thing. "They go to work for Anheuser-Busch."

Jack MacDonough, the vice president for brand management, met them, along with representatives from Budweiser and Bud Light. James Sebo, product manager for Michelob Light, was called in just before the presentation began.

"So that's how you pronounce it," MacDonough said, after the Daedalus group finished its slide show and discussion of the research program. "Now, why don't you tell us how you selected Anheuser-Busch?"

John explained the Seamans–McDonnell–Busch connections, and MacDonough cautioned them not to rely too much on the personal enthusiasms of August Busch for their funding. Of course, it was true Busch enjoyed flying. "He flies everything he can get his hands on," MacDonough said. "A helicopter to work every day, the company jet, even an F-15."

But precisely because of his passion, Busch was particularly careful not to support aviation-related promotional gimmicks. The marketing departments evaluated them in a cold, hard light.

MacDonough seemed familiar with DuPont's success with the Albatross, though, and considered it a good example of how Daedalus might work as an advertising vehicle. "But we're not as modest as DuPont," he said. "We won't settle for a tasteful little sticker on the side of the cockpit." John felt a muscle twitch in his neck.

They talked about turning the flight into a contest, announcing a challenge—like a Kremer Prize—for the HPA that flew the furthest. Make it $1,000 a mile, they said, for the final, record-setting flight of the Daedalus prototype during Phase 2. Well, John said, Paul MacCready might take us up on that challenge. What a stupid idea, he thought.

Then the Anheuser-Busch team wanted to know how versatile the airplane would be compared to, say, a hot air balloon, in case they wanted to use it for public relations events. "It's not the same class vehicle," John said. "An HPA would make a much more impressive static display than a balloon ever would, especially if we set up an ergometer/flight simulator next to it for people to fly. But I don't think we'd be interested in half-times at football games or exhibits in shopping malls."

Okay, MacDonough said. "We have horses for that."

It seemed to John that nobody was talking but MacDonough, and the honchos from Michelob and Budweiser's two divisions hung on every word their boss said.

"Do you think this is a project for Budweiser?" he asked them.

The Budweiser rep paused. "Well, yes, on the one hand, it could be. . . . But, then again, no because . . . Maybe. . . ."

The Bud Light man did the same thing.

When MacDonough turned to Sebo, his Michelob Light manager, John figured the signals must have been clear. "Yessir," Sebo said. "This is just the right market."

This is our guardian angel, John thought. Michelob Light.

Apparently, the turning point of their presentation had come during Ethan's explanation of the physiological aspects, when he'd mentioned that Lois McCallin had completed a four-hour test. Michelob Light, John realized, had targeted a female audience for its beer that just might like to see a woman in the pilot's seat. After they listened to Sebo, Duff began to emphasize the potential upscale audience for the cultural and scientific dimensions of the project. John followed, shamelessly, stressing Lois's involvement as their pilot. In truth, until that point Lois had only been their best test for an engine and the real drive to find better athletes hadn't even begun. But now she seemed not just good, she was perfect.

John told Ethan and Duff later that he'd felt a little like Faust making a pact with the Devil, especially after MacDonough insisted that the prototype airplane couldn't be called Daedalus if Busch agreed to sponsor it. They'd have to call it the Bud Light Bird or Michelob Light Falcon or something like that. John had been so delighted to get a nibble he didn't sniff at their presumptions. He'd even made a counter offer: "Call it the Spirits of St. Louis!"

On the flight home, he wondered aloud about his scruples.

"I guess we'll find out how much our pride's worth," he told them. "Can you imagine us announcing our sponsors in Washington next month—MIT and a beer company?"

Two days before the press conference in Washington, Anheuser-Busch agreed to sponsor the flights of the Daedalus prototype. For an initial investment of $100,000, the marketing managers played their first option and named the airplane the Michelob Light Eagle. They also insisted that John's team would have to provide a scale model of the airplane for public relations events. The Michelob Light logo would appear on the wings and fuselage, the pilot's helmet and uniform, and on the flight simulator. Team members would have to agree to make public appearances with the airplane from time to time. Anheuser-Busch wanted the right to consider adding its other products as secondary sponsors—Saratoga water as the official team water, Eagle Snacks as the official snack. They also demanded indemnification from insurance problems in case Lois crashed the airplane. About the only thing not determined, it seemed, was whether the MIT Faculty Club would have to start serving domestic beers.

No one complained, though. Juan, who'd just finished his humanities requirements and could now claim to be an officially unemployed MIT alumnus, went on the payroll first, at $18,000 a year. Parky, who still hadn't earned a bachelor's degree at MIT, started arranging a master's program between Stanford University, near his home in San Jose, and MIT in Cambridge, where he'd return during the summer to build the prototype. Gup, who was in the midst of a series of airfoil tests for Lockheed in Ohio, couldn't make it to the press conference, but he stayed up all night to finish the details on a one-twelfth scale model of the airplane. He sent it off in pieces, sans fuselage, and when the model arrived at the Smithsonian the day before the press conference, John opened the box and gasped.

He called Mark and told him to get down to Washington immediately to help him finish building the airplane. Then he drove out to a Fairfax County hobby shop to buy glue, epoxy, and even a tiny plastic doll to squeeze into the pilot's seat. John, Parky, and Mark spent a long night in a friend's basement shop completing the assembly. About 11:00 p.m. John wedged a carefully molded, shiny aluminum tube into the figure's hand, meaning to pull it out before the cockpit was sealed in plastic. He was horrified the next morning to see it still there, neatly glued and displayed. "If anyone asks, it's a control throttle," he warned Parky. "Not a beer can, a control throttle."

The flight simulator wasn't originally supposed to be ready for the press conference. But Juan's 102-foot spar wasn't done, and when Anheuser-Busch called, the simulator became a high priority for p.r. purposes. Steve Finberg called a colleague at the Draper Lab, Bryan Sullivan, and the two of them struggled to build something like a video game that would resemble the flight characteristics of the HPA prototype. Bryan had first seen what he wanted in the Artificial Intelligence Lab at MIT. Over in a corner of the lab he'd come across a computer called the IRIS, a machine that was supposed to be used only for building integrated circuits, but that had become most popular because it contained a Silicon Graphics program that let people fly F-16s and 747s without ever leaving the room. By using Gup's QUADPAN aerodynamics program and Parky's final design numbers from the optimization program, Bryan decided he could write all the flight dynamics of the would-be airplane into the Silicon Graphics program. Low and slow, the computerized version of the Michelob Light Eagle answered to the control authority of a joy stick and power generated by cycling on a recumbent ergometer.

Although Bryan and Finberg both worked at Draper, Bryan had no idea what to expect from his colleague in terms of craftsmanship or partnership. He learned soon enough. Finberg did not work nine to five, like many of the engineers at the Charles Stark Draper Laboratory, but instead ran his own graveyard shift that kept him out of everybody's way. He often had a little radio transmitter with him, which he could use to call any number in the Boston area. And he was always in a hurry, as if he were late for an appointment somewhere, trailing a line of cable, stumbling under the weight of batteries, connectors, and tools jostling in his blue-jeans' pockets. People in the working group who had known Finberg since the Chrysalis project said the cafeteria workers at MIT hated to see him come to dinner—he hit the doors at 6:59 every weekday, precisely one minute before the serving line closed.

Two days before the press conference, Finberg was still working on his greatest contribution to the flight simulator. He'd built the joy stick for aileron and rudder control, no problem. But Finberg really wanted to develop a load box—a system for the pedals so they'd grip with torque, to simulate the dynamics of the propeller when the pilot strained for more power. He built an automobile generator that dynamically simulated the load of the airplane, and then went to work on ways to increase the pressure when the pilot pedaled faster. Bryan assured him that dynamic loading was not necessary for the simulator's first public demonstration in Washington. But Finberg was convinced he could get it working.

They were supposed to test Finberg's "load box" at MIT the night before Bryan left for Washington. The team had arranged to borrow the IRIS and drive it down to D.C. the day before the press conference. Bryan and Finberg had volunteered to set up their device at the Air and Space Museum. Except for the load box, the Silicon Graphics program was ready to run.

"I'll be there at midnight," Finberg told Bryan, the afternoon before the test. Bryan and Juan waited at MIT until midnight. No Finberg.

At 2:00 a.m. Bryan called Finberg at his office. "Yeah, yeah, it's almost done," Finberg said. Bryan took a nap, and at 4:00 a.m. he woke up. No Finberg.

"Yeah, I'll get it there," Finberg said when Bryan called again. At 6:00 a.m., still no Finberg.

Bryan, who was scheduled to be at the museum with the flight simulator at 10:00 a.m., caught the next flight to D.C. Juan agreed to pick up Finberg and the load box in the aero department's van, and drive them down to Washington. At mid-morning, he found Finberg at the Draper Lab, where the load box lay in a clump of disconnected pieces. They put everything in boxes, climbed in the van, and headed out for the highway. "Wait," Finberg said, "you've gotta go back. I forgot my suit."

Juan took Finberg home to get this suit. But Finberg revered his privacy so much that no one from the working group had ever been to his home before—or even knew where it was. So before giving Juan directions, Finberg made him promise first not to reveal its location to anyone. Juan promised.

As they finally nosed up to the turnpike entrance in Cambridge, Juan tossed a quarter into the toll machine and stepped on the accelerator.

"Wait!" Finberg shouted. "Stop!"

Juan hit the brakes. "What? What is it? What happened?"

"Get a receipt."

"Finberg!"

When they arrived in Washington, it was late afternoon, less than twenty-four hours before the press conference. Debbie Douglas, an historian still on loan to the team from the Smithsonian, rushed Finberg upstairs into the museum, past the security guards, and told him to hide there until the museum closed. He could use her office to finish building the load box.

At 4:30 p.m. Bryan got a call on a little radio receiver Finberg had left with him.

"W1GSL, W1GSL," it barked.

"Yeah, Finberg," Bryan answered. "Is the load box ready?"

"Do you guys know where I could get some capacitors?"

Bryan called John's wife and got directions to the closest Radio Shack.

Five hours later, another call. "W1GSL, W1GSL."

"Yeah, Finberg," Bryan answered.

"I've gotta get some solder. Do any of you guys have solder?"

Bryan found solder at an all-night electronics shop in downtown D.C. By midnight, he and Juan were finished with the flight simulator and left to get some sleep. Finberg remained secluded upstairs in a museum office strewn with generator parts.

"So, you've had your first taste of Finberg?" Juan said on the way to their hotel.

"I'm no longer a virgin," Bryan replied.

On April 9, 1986, at 9:00 a.m., MIT, the Smithsonian, and Anheuser-Busch announced their sponsorship of the Daedalus Project. The press conference went off without a hitch. *The Washington Post* published a lengthy article about the project on page three. Finberg's load box worked perfectly with Bryan's flight simulator and the ergometer. No one even asked about the beer can in the pilot's seat. Members of the team walked over to the Capitol to have their pictures taken, and only Ethan puzzled aloud about this peculiar mix between MIT and the brewery. "I wonder what Yale will think about this," he said, as cameras snapped.

In fact, the only thing to happen that day that really seemed unexpected was the sudden appearance of a public relations official from Anheuser-Busch who flew in that morning with miniature markings for the scale model. Just before the press conference began, John watched the guy, dressed in a natty three-piece suit, carefully and quietly apply Michelob Light Eagle decals to the model airplane.

This is interesting, John thought. Why are they going to all this expense just to put a couple of decals on the model?

He hadn't quite figured out yet that it wasn't just their airplane project anymore. It would scarcely be their airplane at all.

12

Guests traveled an impressive circuit when they visited Hanscom Air Force Base in mid-July. The Michelob logo appeared in every photographer's viewfinder. Journalists groped through an impressive morass of technical documents and looked at project sites strewn over twenty square miles of suburban Boston.

They'd nose through the woods between Lexington and Concord and, taking a winding rural route to the hangar, find a secluded cinderblock building that housed a dozen top-secret research projects. A security guard and a dog named Gyro would meet them at the door, then a project guide would lead them across the cool concrete floor to a corner, where the Daedalus prototype was taking shape. Outside, in a rust-colored shed, student volunteers, who snoozed in sleeping bags and snacked on potato chips and Cokes, worked more than one hundred hours a week and made about eighty cents an hour. Just beyond the shed, across a flat, treeless field, was the runway where they'd test the plane.

Sixty miles south, several thousand feet overhead, Bussolari coached Lois in a fussy sailplane above Plymouth Airport. Fourteen miles east, Mark and Parky meticulously ground slivers of steel from a gear. Mark's fingers would be slippery with oil while he worked to cut a few ounces off the propeller's gears, and Parky would boast to visitors that the plane would weigh only sixty-eight pounds when they were done. The Eagle, he'd explain, will have a wingspan wider than a DC-9 jetliner.

John greeted guests and guided them around like a ringmaster. He showed them how the team mixed epoxy to join the spars. Sweating in the humid air that stirred through the resin-scented hangar, he punched sticky fingers into a calculator for their instruction: eleven grams of resin required three grams of epoxy catalyst. "Sort of tough to tell it's an airplane yet," John said. He weighed the goos, mixed them, looked at his watch. Twelve minutes before the glue dries; 6,000 hours before the Eagle flies.

But even professional observers couldn't distinguish industry from mayhem. Press coverage usually stressed the "Whiz Kid" angle. The timetables they saw said the airplane would fly by August, and no one disputed the manager's bold estimates.

In truth, behind the facade of organization and efficiency, the project eked along months behind schedule. The dozen or so undergraduates who'd signed on to build the airplane (they were called UROPs, for the Undergraduate Research Opportunities Program) typically knew nothing about even the simplest workshop techniques. The UROP assigned to help Juan make carbon spars confessed that she'd never held a hand drill before. The boy who volunteered to make a set of drawings of the elevator spent $200 on drafting equipment but he had no idea how to use it. The most experienced UROP among them built mandolins, but he couldn't machine aluminum hinges for wing mounts.

The undergraduates idealized Parky and tolerated Juan's bad moods. As disorganized as they were, everyone still had a chance to build significant pieces of what could become a historic airplane. It was tedious, time-consuming, repetitive, and boring work, but the people milling around the construction sites were creative and intelligent. And the more time they spent at Hanscom, the more it felt like it was their airplane, too.

Poor Juan wasn't meant to be a supervisor so early in his career. He worked endlessly. He complained it was the worst summer of his life. From 10:00 a.m. to 3:00 a.m. he wrapped spars and worried about weights. Despite warnings from MIT faculty, the idea of building spars with three carbon tubes (the "three kissing tubes" notion) turned into a vast time sink. According to the schedule, Juan should have finished the spars by June, but by July he and two UROPs were still wrapping mandrels and testing parts. Worst of all, the particular composites he'd ordered weren't available, and Juan had to settle for a heavier grade of carbon fiber. The estimated weight of the airplane rose from sixty to eighty pounds, and Juan blamed himself, privately, for jeopardizing the project. Making Daedalus became an obsession. He studied mythology and worked, increasingly, all night long after the UROPs had left.

John himself was not the easiest of taskmasters. He'd moved Barbara and their son, Ellis, to Boston to live at a friend's house. But the temporary arrangements soon palled. Not knowing whether the project would survive past the summer, Barbara and Ellis shuttled back and forth from Boston to the family's home on Cape Cod. Finally, they just stayed on the Cape, and John went to live at Jack Kerrebrock's house. He saw his family only once a week—on Sundays—in between haggling with Juan about spar delays.

John expected Juan to provide leadership for the undergraduate builders. He was also concerned that weight increases in Juan's carbon tubes could make the airplane too heavy for Lois to power. He'd look at Ethan's data and then at his own, but he could never bring himself to do the analysis. Lois is doing great, he'd think. And even if these numbers are right, what am I going to say? "You're dead. Forget it, honey"? He half-hoped Ethan would find another engine.

With no one else to blame, John rode herd over Juan. When Michelob sent a set of specially prepared Mylar decals, made out of 1-mil rather than .5-mil plastic, Juan carefully took a razor to the material and cut out every pinch of unnecessary Mylar from around the letters. Just a precaution, he explained. Saves weight. If this is some kind of joke, John thought, it's not funny. He prodded Juan incessantly to stick to the schedule. And more than once, Juan thought about quitting.

But in truth, they all shared the burden of guilt for the delays. The final design of the airplane, for instance, changed weekly. The pusher prop drawings, which had appeared in the final working group report at the press conference, went into the trashcan as soon as Parky showed up for work in June. "Gee," he said, looking at the drawings, "I leave you guys for a couple of months and you do this to the airplane?" In all previous HPAs built with pusher props, the drive train system, which used a twisted bicycle chain, had created serious malfunctions. "Don't you remember the Albatross team T-shirt?" he asked. " 'If the wind don't blow and the chain don't break. . . .' " Parky got rid of the chain and redesigned the airplane with the propeller in front. Delay: approximately three weeks.

For several weeks, too, the fuselage design changed daily. Every morning different members of the team pinned an improved scheme for a streamlined cockpit on the door of the mezzanine in Building 33. The "fuselage of the day" competition continued until they settled for a cockpit that looked like an aristocratic nose—thin, highly swept back, with discreet little air ducts. Delay: approximately one month.

Mark's airfoil design was exceptionally difficult to execute. Because of their contours, the wings required three separate sections. The dramatic taper from the center panel to the tips demanded that every one of the one hundred fourteen Styrofoam ribs supporting the airfoil have a slightly different shape, varying as little as a couple of millimeters from one another, up and down the spar. Juan's spars also had to conform to the design, so instead of simply making circular pieces, he had to devise a mandrel with a slightly conical shape. Delay: approximately two months.

Eight of them eventually moved into Jack Kerrebrock's house so they could be near Hanscom, increase their efficiency at work, and conserve energy. Jack and his wife, Vickie, never knew who was there, since most of the team lived on an inverted schedule and their paths rarely crossed. Vickie would put out breakfast cereals, eggs, and juice for them every morning before she left for work, and some days she'd return home to find a cot laid out in the playroom and a bundle of strange clothes in the dryer. They interacted most often through the icebox.

Parky, who was taking courses to finish his undergraduate requirements, established a radical routine: he rose every morning at nine, drove into Cambridge for his ten o'clock class, went to work in Don Wiener's lab when it closed at 5:00 p.m., machined aluminum and steel parts for the gearboxes until 11:00 p.m., called his wife in San Jose, kept working until 2:00 a.m., drove back to the Kerrebrocks', took a shower, and fell into bed about three. Every other Thursday, Parky caught the five o'clock flight to San Jose, crossed his front stoop about 9:00 p.m., spent Saturday in bed sleeping, read science fiction until three o'clock Sunday, spent a few hours with his wife on Sunday afternoon, and at 6:00 a.m. Monday caught a flight to Boston, where he rose at nine Tuesday so he could make his ten o'clock class at MIT . . .

Mark, who'd joined the MIT faculty, found that his new academic obligations cut into the precious time he had for building the airplane. The solution to the twenty-four-hour day, he decided, was to steal a few hours from the seventh day and live on a six-day week. Since he had a faculty meeting every Monday morning, he'd start his week then. Every day he added a couple of hours to his schedule and worked a little longer. By Saturday night his biological clock was set to keep him awake in the lab all night. Sunday he'd climb in bed early enough to make the faculty meeting Monday morning. He lived on twenty-eight-hour days. His mother would call from Philadelphia and find him at work in his office on Saturday nights as late as 1:00 or 2:00 a.m.

"Mark," she'd say, "do you have a girlfriend? You have to find a girlfriend. . . ."

Somehow they all managed to get along well enough. Excessively hard work, sneering, and friendly taunts became a part of the ethos. Around the lab, the shop, and the hangar, anyone who took a break or tried to sneak out early risked being called "slacker." During work weeks that ranged from eighty to 100 hours, "slacker" turned into an effective epithet and set the standard for commitment and energy. How hard could they work? If only time would tell, it was exceedingly clear, they weren't working hard enough. After just three months, the project was perilously close to failure. It should have been an awful time. But compelled by the dream of Daedalus, the team's day-to-day experience of independence and spontaneous engineering generated all the motivation necessary to keep the project alive. After all, what else was there?

They'd gather around a drafting table and Mark would pull out a drawing of his wings lightly sketched on a sheet of drafting paper. John flicked a pencil between his fingers, Parky clenched a pen between his teeth, Juan flipped a mechanical pencil end-over-end into the air. They would joust.

"Okay, how do we do these ribs?" Juan asked.

"You mean, how do we build them?" Mark said.

"I mean, these ribs have to take the tension of the Mylar covering, so unless there's some structure there, we're going to warp the hell out of the wings. If we just put some balsa strips on the face of each . . ."

"Wait, wait, wait," Mark said. "This is too complicated. Forget these things."

"Mark, these things are really important."

Then they'd fight. The participants would lean over to pencil in a design change on Mark's drawing. Theoretically, they'd already built the airplane. But when it came time to make real parts, a mass of problems gathered like stormclouds over their computer printouts: how do you keep Styrofoam from buckling; what's the most efficient way to lace Kevlar twine for bracing the internal structure; what combination of materials will make strong wing ribs? They'd argue for a half hour, then they'd start again.

"I mean, how beefy should these ribs be?" Juan asked.

"They don't have to be beefy," Mark said.

They'd argue about what kind of material to use for the leading edge—blue foam, Styrofoam, Rohacell, fiberglass—then they'd go back to the original problem, how to make the ribs.

"On Monarch," Mark said, "we had a tube that butted against the carbon side strips on the ribs . . ."

"Yeah, I know," Juan said. "Then it had a little wooden block beside that, and it sucked!"

"No, it worked fine. What are you talking about?"

Juan pointed his finger at Mark's head, cocked his thumb, and fired. Ka-booom! "Well, that was pretty bad," Juan said. "I put it together and it was gross and it was really inadequate."

"We're talking about the Monarch's beefy ribs?" Mark asked.

"Yeah," Juan said, "I built that."

"Well, I don't remember it."

"It doesn't matter, Mark," Juan said. "It was just basically gross."

"Okay, do whatever you like."

"So," Juan said, "what we can do is bury a graphite tube inside the rib . . ."

"Great!" Parky said. "The worst of both worlds."

The question of how to build one hundred fourteen differently sized ribs and one hundred two feet of finely tapered Styrofoam sheeting for the leading edge stymied the team. Given six months, perhaps Mark could make the cuts. After working with ultra-lightweight gliders for so many years, Mark had the keenest sense of proportions of anyone on the team—he could distinguish between a ⅜-inch drill bit and a ¼-inch drill bit from across the shop floor. Mark, they said, had cross hairs in his eyeballs. But Mark didn't have time to shape the wings. The best idea they'd had was to use

Mark's ISES airfoil program to trace the outline of each rib and then make computer-generated sketches to cut one hundred fourteen Masonite templates for the Styrofoam ribs.

They still didn't know how to cut the curves for the thin Styrofoam fabric for the leading edge that would clothe the ribs. They hadn't a notion—"clueless" is how Juan would put it. The word rolled off their tongues like an emptying sack of marbles. The computer could draw contours, but no human hand was steady enough to guide Parky's hot-wire foam cutter. Manufacturing the leading edge required extraordinary precision.

Steve Finberg watched Mark test a few Masonite templates in the lab one afternoon and listened as he and Juan groused.

"Sounds like a bear," Finberg said.

"Yeah," Mark said, "We've gotta make one hundred fourteen different wing sections, too."

"That would take approximately one hundred UROPS maybe two months," Finberg said. "I'll bet I could build something. . . ."

Mark and Finberg envisioned a machine fourteen inches high, sixty inches wide, and four feet long that would cut the wing ribs and leading edge sheets without human interference. John called it CAD-CAM, computer-aided design, computer-aided manufacture. And Juan always objected. "*K*-A-D, *K*-A-M," he'd say. "Knowledge-aided design. Knowledge-aided manufacture." No matter. The theoretical designs that began in Mark's imagination would be ground into perfect form by the Perkin-Elmer computer, translated for numerical machining by a software program Mark devised, fed into the brains of a remarkable cutting machine Finberg would build, and brought to life by a thin hot wire dragged through a block of Styrofoam. There was just one problem: only Finberg could build such a device, and Finberg, as everyone knew, had a rangy sense of time.

After a couple of months as Finberg's electronics assistant, Bryan Sullivan lost his patience. It was a Saturday night in mid-July. The first wing rib hadn't been cut. The leading edge was still a two-dimensional piece in Mark's computer. The UROPs wasted time playing pilot on the flight simulator and Finberg was late getting to the lab, as usual. About 9:00 p.m., Finberg strolled in.

"Good morning, how are you?" Bryan said in a whiny sing-song. He realized it would be another all-nighter.

Finberg fiddled with a few boxes of parts, the oscilloscope and circuit boards, then puzzled silently over Bryan's drawings of their foam cutter design. After an hour, Bryan understood that there was nothing left for him to do but sit and wait. He picked up a hunk of foam and a razor blade and shaved away the corners.

"What are you doing?" Finberg asked. Bryan was staring.

"I'm carving your face here," Bryan said. "I hope you don't mind."

Finberg nodded and went back to the box of parts.

About 3:00 a.m. Bryan went home. Before he went to bed he finished shaping the Styrofoam, cut a little stretch of aluminum wire to make a miniature pair of glasses, and gathered some cotton balls to fluff into a beard. The next morning he carried the finished piece back to the lab and left it sitting on top of Finberg's pile of parts. When Finberg came to work Sunday night he found a Styrofoam head, which looked remarkably like his own, shrunken down to about one-quarter scale. A note taped to the bald pate said: "Either this hot wire works or it's your head, Finberg."

The next week Finberg called everyone in the shop up to the mezzanine to see his machine. When they edged up to the table, they smelled the last acrid puffs of smoke from a steaming soldering gun.

"Meanwhile," Juan said. "Weeks later. . . ."

Finberg grabbed a pair of needle-nosed pliers and stuck his head between two long rods on the machine to make the final adjustments. "Oh boy, this is impossible to get at," he said. "If it just doesn't decapitate me first. . . ."

Bryan, John, Juan, Parky, and several UROPs gathered around. They looked haggard from lack of sleep. Juan's hair hung down past his shoulders and he stared out from under the white brim of a Michelob Light cap. Not even Bryan, who'd worked for two months on the machine, believed it would work.

Finberg flipped a switch on the back of the device and tapped a white button above it. The rods whirled and a set of gears the size of half dollars revolved, animating a series of chains. He stepped over to an IBM personal computer stationed on a desk nearby and tapped a few keys. "Sooner or later, I'm just going to start dumping Mark's files into it and see what it does. Want to see how many times I can iterate?"

They watched dubiously. The wire did not move.

"These things are slow," Finberg explained.

They watched for another minute. The wire did not move.

"Well, not that slow," Finberg said. "Bryan, where's the schematic?" There were a dozen empty Coke bottles laying around the tables. Finberg pulled a circuit board out of the machine and reviewed the drawings. "Okay, let's try again," he said. "Hopefully it's steady enough. It's steadier than you could drag it by hand." He flipped the switch and punched the button. The wire flinched.

"John!" Finberg shouted.

"Let's get a piece of foam," Juan said. He grabbed a hunk of white beadboard from a shelf.

"You mean, it's never cut a piece of foam before?" John asked.

"Foam?" Finberg replied.

Parky folded his arms and watched Juan clamp the Styrofoam onto the machine. Finberg tapped a couple of keys on the IBM, and very slowly the rods began to drag the hot wire through the white board.

"I hope it will go faster than this," Juan said.

"You notice how it pauses every tenth of an inch for twenty milliseconds or so?" Finberg said. "Well, I've eliminated that in the final design."

"We don't want the pause," Parky said.

As they argued, the wire continued to bite through the hunk, forming an airfoil that matched the precise dimensions prescribed by Mark's computer design.

"Does it ring some alarm when it's done to wake you up?" John said. "It's pretty boring. I mean, it's pretty neat. Actually it's awesome."

"Does that mean we haven't totally lost the last three weeks?" Juan asked.

The first test piece to emerge from the teeth of the foam cutter was a stairstep block about four inches high and four inches wide. The second piece was a cruciform, approximately four inches long. A few days after the machine's inauguration, someone noticed the pieces, redesigned in architectural fashion, resting on a shelf. The stairsteps led up to a small platform, where the cruciform had been glued upright. Plugged into the peak of the cruciform was the Finberg head. It was labeled, in Magic Marker, IDOLBERG.

A new foam cutter was born, and with it, a myth—the myth of Finberg.

The foam cutter really was a remarkable tool, as sophisticated as anything used in the airplane industry. There had been no blueprints, no drawings except for the circuit designs, no mistakes except in the computer, where construction time was instantaneous. It was a wonderful invention. But no one cheered its arrival. They didn't have time to pat each other on the back. They couldn't afford to celebrate. "This isn't a support group," John said.

But they did retain the Finberg head. It was said that the Styrofoam object had restorative powers, and from that day forward the head was passed between team members whenever trouble brewed or a bit of good fortune was required. The new foam cutter advanced the project forward by several months.

Good publicity brought them luck and momentum. A Greek physiologist visiting at Harvard University read about the project and offered his services. Within weeks Steve Bussolari and Ethan received invitations to the Greek Olympic Training Center in Athens to conduct VO_2 Max tests on the Greek Olympic cycling team. At Yale, a series of outstanding cyclists slipped into Ethan's lab to try out as pilots, including John Howard, who'd recently broken the world speed record for cycling at 152 miles per hour. A Norwegian company, whose president was an alumnus of MIT, donated a valuable set of weather instruments, and Parky and Bussolari flew to Greece to install the machines at points along a potential western Daedalus route from Akra Spatha to Maleme.

But they still didn't have enough money to make test flights for Anheuser-Busch. Searching for flat, dramatic, windless open spaces, they discussed flying over a frozen lake in North Dakota; they sent a UROP down to Florida to investigate a route over the Everglades. At night John desperately thumbed through financial statements. The Anheuser-Busch donation would serve them through the construction effort, but there wasn't a cent to pay for the flight of their prototype. As hard as he'd looked, no other company in America wanted to donate money to an airplane called the Michelob Light Eagle. The billboard was full.

While the team struggled to advance on schedule, John played with the balance sheet to see if he could actually save money by breaking away from MIT. If he formed his own private business, John suspected he could dump the high overhead costs charged by the university for the use of its facilities, its staff, and its professors' time, as well as cut some of the employee benefits that seemed least necessary. The costs for carrying a private venture always came up about 30 percent less. He fantasized about severing ties; whenever people spoke of his team as an MIT project, John began to feel a twinge of resentment.

The budget fell $100,000 short. They had until September to make initial test flights, until October to break the twenty-two-mile distance record set by the Gossamer Albatross, and until June 1987 to build the actual Daedalus airplane and fly across the Aegean. Delays dogged them. At night even John joined UROPs making ribs and stringing Kevlar twine through the wings.

Late in the summer, Parky sculpted a couple of sets of gearbox casings out of a block of forty-pound aluminum. In the end, the upper gearbox weighed about a pound and a half, contained eighteen machined parts, eight ball bearings, two gears, and more than one hundred screws, some of which Parky carved by hand. He'd turned the boxes over and over in his hands thousands of times, trying to comprehend the puzzle of gear ratios, torque, and angles.

The boxes had to be light and strong, able to withstand up to six hundred pounds of torque from the pilot's pedaling and not deform by more than 1 millimeter in the gears. Even the bearings had to be made by hand. The gears themselves, donated by the Arrow Gear Company in Downers Grove, Illinois, fell under Mark and Parky's lathe; the gears were customized to shave off half their weight. And because the loads were so severe, Parky had to shrink fit the gears to the shafts, rather than using pins to hold them in place. In the end, the angle between the shafts had to fall between 90 degrees and 90 degrees, 2 minutes. The gears of a Swiss watch might be tighter, Parky said, but then a Swiss watch doesn't have to handle six hundred pounds of torque.

Juan built more than forty carbon tubes, with the help of his UROPs. The undergraduates became more efficient and began to take on tasks independently. Juan's mood didn't lighten all summer, but he hit the deadlines for construction, and a few members of his crew developed a loyalty to the team's only structures expert. Juan turned out to be an excellent teacher.

Everyone talked about "the learning curve." After wasting hundreds of dollars in materials to make trailing edges and wrap Styrofoam ribs with balsa strips, the UROPs were finally working without supervision. The engineers' basic quality-control mechanism—TLAR (That Looks About Right)—began to make sense to them. Perfection mattered less, and the intuitive notion of what was good enough became clear. Tom Clancy, an electrical engineering student whose father was a sculptor in New York City, began to view the airplane as a work of art. Grant Schaffner, an aeronautics major from South Africa, calculated the proper spacing for the ribs and learned, at Parky's side, how to mold a fuselage using a Styrofoam plug and fiberglass.

They tested the prop one night at Hanscom near the end of the summer. Mark mounted his twelve-foot-long prop on a wingless carbon frame, knelt on the floor, and arched an eyebrow at the vertical axis. For greatest efficiency the prop had to be 1.25 degrees off the perfect vertical angle. Mark made a few adjustments in Parky's upper gearbox, and climbed into the pilot's seat. The propeller spun at one and a half times the pedal speed.

It had been shaped like no other propeller in the world—accounting for all the viscous effects of drag on the airfoil. From tip to tip it was the optimum design, made of lightweight foam for bulk, covered with Kevlar cloth for strength. As Mark cranked the pedals, John and Parky and Juan felt a strong wind wash from the propeller, tousling their hair, blowing Kevlar threads on a sawhorse thirty feet away almost horizontal. "Actually," Mark said, increasing his pedaling rate, "you should be able to feel the thrust about now." It took three of them to keep the frame from scooting across the concrete floor.

By the end of September, as pieces of an airplane became visible at the Hanscom hangar, strong emotional attachments held the team together. They spoke of themselves as a family. No one referred to their airplane as the Michelob Light Eagle—unless they were talking to the press; instead, they called her Emily, born from the acronym MLE. A solar-heated shed near the runway became the cradle of wings, spars, fuselage parts, control panels. There they ate breakfast, snacked on M&Ms and Cokes, slept. The smell of dozens of different kinds of epoxies and compounds hung in the air. Parky called it a toxic waste dump. But it was also a home, of sorts. It was there someone first clamped a hockey puck and a lawnmower wheel to the fuselage so the airplane would roll across the runway. It was there, one day, that Lois appeared wearing a Bumblebee costume. And although most of the UROPs disappeared in September to go back to school, a couple, like Tom Clancy, went ballistic and quit attending classes completely so they could stay at the "solar home" and finish building Emily.

By summer's end, though the airplane still existed only in parts, the engineers carefully pieced together what they had, and John invited representatives from Anheuser-Busch out to Hanscom for a program review. A film crew from NBC showed up and a reporter from *USA Today* made an appearance. Though the airplane was far from completed, Daedalus team members pretended that they'd made tremendous progress. John spoke glibly and the Anheuser-Busch guys believed every word. Increased funding followed: another $100,000.

After announcing a series of "death pacts," the team mounted a nonstop campaign to finish Emily. By October they began test flights at Hanscom, and John told NASA officials at Edwards Air Force Base in California that his team wanted to attempt its record flight in January 1987 on the dry lakebed at the Dryden Flight Research Facility. The actual Daedalus flight would have to be delayed for another year, and the issue of further funding could be avoided a while longer. But at last they had an airplane, and they boasted to the world that they were ready to challenge MacCready's record.

On Wednesday, October 15, the new Michelob Light Eagle rolled into the Hanscom hangar and John addressed more than one hundred guests: "The prime goal at Dryden will be to accurately measure the power and pilot cooling requirements, but the milestone that will attract the most interest is the attempt to break the existing world record for distance. For those of you who wish to see a much shorter flight, we will try to fly here at Hanscom tomorrow and Friday. If you are accompanied by a team member or have press credentials, you may watch from the designated area on the field. Otherwise, the best place to view the flights will be from the side of Virginia Road on the grassy hill there."

Emily flew the next day—short hops up and down the runway. Casual spectators, press, even some faculty members who showed up, were easily taken in. The airplane might have made it across the Hanscom pavement, but it couldn't fly for more than three minutes. It never had. Lois had just enough power to keep it afloat for the benefit of photographers and journalists. But the real story appeared in the weight charts:

ITEM	PREDICTED WEIGHT	ACTUAL WEIGHT
body	32.23	31.47
spar	17.36	32
Leading Edge	2.62	6
ribs	2.24	4.6
covering	1.6	2.5
Trailing Edge	2.55	3.0
wire	0.408	1.0
tails	1.12	3.8
misc.	0.0	2.7
Total pounds:	60.13	87.07

At eighty-seven pounds, Emily squatted down on the runway like a constipated ostrich. In the air, she was a flying pig. John smiled through the whole affair. They said he did a nice job on Good Morning America. He gave a good sound bite and snappy quotes. After all the favorable publicity, not even the public relations guys from Anheuser-Busch suspected that the whiz kids really couldn't fly.

13

John never blamed Lois. She'd made every four o'clock wake-up call for flight ops, and each week devoted herself to a disciplined training regimen, running and cycling more than two hundred thirty-five miles through the wooded back roads around Hanscom. She'd even made a special effort to learn aerodynamics so she could carry on conversations with journalists who called wanting to know how in hell someone could pedal a Styrofoam airplane seventy miles. She could talk Reynolds Numbers. She could talk boundary layer. John watched her all-out sprints down runway 5-23, and chatted with her in the cockpit as she heaved and sweated on frosty autumn mornings at the end of exhausting two-and-a-half-minute flights. He could see the devotion in her face. Lean and bony, Lois almost swaggered when she walked. She was so quiet and determined. She'd made great progress as a pilot during flying lessons with Bussolari. The engineers were crazy about her. No, Lois wasn't the problem.

Increasingly, what angered John was the rest of his team's cavalier attitude. He noticed a couple of young UROPs spending more time fawning over the pretty engine than shaving ounces off the airplane. Entertaining Lois. Flirting with Lois. Making dates with Lois—or that's what it looked like to him, anyway. He discovered builders swilling Michelob gratuities in the afternoons, and bristled when they complained about the Eagle Snacks running out. The airplane flew so irregularly that they couldn't even gather enough data to estimate how much more power they'd need. His chief engineers came up with a dozen excuses for the airplane's sluggishness and dallied over actually refining Emily.

Be calm, they'd tell him. Don't worry, we'll get it done. They laughed behind his back and mocked his agitated expression—"the trembling lip," they called it. As he spent an increasing amount of time in an office, arranging logistics and talking to the press, John's own reputation as an engineer faltered among his peers. They labeled him a "management type," and questioned his craftsmanship. If it's not on his schedule, it doesn't exist, they'd say. Juan and Parky played games to keep John away from their airplane. In his haste, they worried, the manager would break a wing.

Even the physiologists seemed nonchalant to John. He wanted data on temperature regulation of the cockpit, but he never got what he considered adequate responses from Ethan or Bussolari. And after several months, they still seemed to have made little progress in winnowing out another suitable engine.

As many as a dozen athletes had tested in the lab throughout the summer, including John Howard, the fastest cyclist in the world, and another guy who'd just won the Race Across America. But the best of the lot had to be discounted for fear that, as John would say, "their egos won't fit into our cockpit." John had spent enough time on the phone haggling with cycling agents to have grown wary. He told Bussolari and Ethan to find him a better engine, one who met three requirements: the pilot had to be self-deprecating enough to allow engineers their rightful place in the spotlight, intelligent enough to inform the press about the particulars of aerodynamics, and trained well enough as an aviator to skip months of private lessons in gliders. They couldn't take a chance with a known personality. Daedalus had been an engineer, after all, not an athlete. They needed an obscure but promising also-ran, who would stand in the shadows and purr like an engine after the flight.

The two professors hadn't discovered a single candidate with high specific power who shared the characteristics of humility, intelligence, and piloting skills. Besides, Ethan wasn't on payroll to find them the perfect engine, and as a member of MIT's junior faculty, Bussolari carried a heavy teaching load. Like the engineers, they neither felt nor completely sympathized with the pressures engendered by the $200,000 promise John made to Anheuser-Busch. A nettling tension was brewing between the manager and his engine builders.

As friction between John and the team worsened, a part-time assistant at Yale University's Pierce Foundation labs heard about the project through some of Ethan's associates. Sometimes, in the afternoons, Glenn Tremml would slip into the mock flight lab to watch famous athletes being tested as potential Daedalus engines. Wow, he thought, John Howard, Rudy Sroka, Jonathan Boyer! To the twenty-six-year-old research assistant, those guys were heroes. They sold sunglasses in cycling magazines, sweats in triathalon periodicals. He was awed by the sudden appearance of regional and national figures in his workplace. He wanted to meet them, and he wanted to know more about the mysterious project.

One afternoon, Glenn and Gary Mack, one of Ethan's assistants, went for their regular ten-mile run through New Haven, and Glenn mentioned that he'd seen some of the VO_2 Max tests for the Daedalus Project. The charts looked really good.

"I'm impressed," Glenn said.

"Yeah, well, the only problem is none of them are pilots," Gary explained. "We're looking for people who are fit, but they've gotta be able to fly."

"Oh?" Glenn said.

A few hours later, Gary Mack appeared excitedly at Ethan's door.

"You know Glenn Tremml?" he said.

"Yeah," Ethan said.

"Well, do you know he does triathalons?"

"Yeah, so?"

"And do you know what else?"

"No, what?"

"He's got a pilot's license."

Glenn's VO_2 Max test astonished the guys in Ethan's lab. At 3.3 watts per kilogram, he tested as well as some of the national-class athletes they'd seen. His self-effacing manner suited the role they wanted to play before the media. As a medical student, he already spoke the language of physiology. And best of all, he was a trained pilot. Glenn Tremml might not have been the perfect engine, but he was by far the best Ethan could produce. By mid-November, just six weeks before Emily was supposed to make a two-hour flight across the mottled floor of Mojave, he offered the team its greatest promise. Lois was still hobbled by the airplane's weight and the engineers had discovered no new solutions for her. John waited on the results of Glenn's four-hour test before calling his crew together. If

Glenn could show enough power in the final engine test, maybe an unpleasant showdown could be avoided.

Sloe-eyed, happy-go-lucky Glenn Tremml was about to join the maze.

Four hours is a hell of a long time, Glenn thought. I hate sitting in the car for four hours.

He turned the time over and over in his mind while he trained. Four hours is a hell of a long time, he'd think, as he ran the rutted paths around the West Hartford reservoir near his home. Isn't that how long it takes to drive to my aunt's house at Christmas . . . ?

As he prepared for the final experiment in Ethan's lab, Glenn analyzed his own anatomy to estimate the chances for success. He understood almost too much about science and the limits of long-endurance exercise. Unlike many athletes, he didn't trust the gut-it-out approach to sporting events. A four-hour ergometer ride sounded like the equivalent of running two back-to-back marathons to him. That's right, Ethan told him. The test wouldn't be easy.

He was an unlikely Daedalus figure. He certainly had the good looks to match the image of a Greek hero—thick, curly brown hair, a patrician nose, an angular chin and sharp cheekbones, deep brown eyes. At the same time, he lacked the verve and confidence of the Greeks. Throughout his life, Glenn's tendency in athletics had been to quit a sport as soon as he realized his limits. As a child he'd failed at Little League baseball, so he'd tried judo, and wasn't particularly successful at that, either. He went from equestrian events to ballroom dancing. When he finally switched to swimming, his standing improved. At fourteen, he ranked first in Connecticut as the state's fastest fifty-yard free-style swimmer. At that point, Glenn decided he could be an Olympian by the time he reached his eighteenth birthday. Between the ages of twelve and eighteen, he swam five or six days a week, training for his chance at national competition. Unfortunately, his body quit developing. While his friends grew tall and gained bulk, Glenn retained the modest frame of a fourteen-year-old. At fourteen he stood five foot nine and weighed one hundred forty-seven pounds. At twenty-six, he stood five foot nine and weighed one hundred fifty pounds.

No amount of preparation—psychological or physiological—could overcome nature's decree.

At Dartmouth, Glenn joined the swim team and competed primarily with the junior varsity—as a diver. He usually finished second. Even at fraternity parties, he couldn't claim local-hero status as a beer drinker—try as he might, he could down a few drafts, but not gallons like the big brothers. By the time he graduated from Dartmouth in 1982, he'd given up childhood's dreams and concentrated on getting into medical school. No more fantasies, he thought.

After three years with the Pierce Foundation, making a dispiriting $9,000 a year as a research assistant, Glenn finally used his experience in the lab as a bargaining chip to get into medical school at the University of Connecticut. (His résumé had grown to include a couple of published scientific papers and, in a curious but not unrelated addendum, a pilot's license: Once, on a trip to Kentucky, where Glenn was going to present a scientific paper, he caught a flight in a little eight-seater shuttle plane, and as the airplane popped in and out of the clouds at 2,000 feet, the excitement of flying rekindled a childhood interest. Two days later, he went to the New Haven airport and signed up for flying lessons.)

By the time he entered medical school, his body had turned flaccid. He saw only one more year left during which he might compete in amateur athletic events. School would absorb all his time by the third year, and after that, he assumed, he'd hit the downhill slide of maturity.

During the previous year, when he should have been studying enzymes in medical school, Glenn started training seriously for triathalons. The summer of '86 was to have been his last effort in athletics, and his workouts became lengthy and exhausting. He grew lean and fiercely competitive.

When he learned about the Daedalus Project in August, Glenn still hadn't won a single triathalon, but he was in excellent physical condition. He may not have been a challenger in the water or on the ground, but with an extraordinarily high aerobic capacity, Glenn soon discovered that he was unusually well suited to the air. Ethan's stats bore it out: his body fat was a low 5.2 percent; he had the projected ability to maintain a heart rate of 165 beats a minute and produce 225 watts of power for several hours without tiring; his anaerobic threshold—the point of physical exhaustion where the body incurs oxygen debt—was exceedingly high. Pound for pound, he stood on par with national-class athletes. Just as astounding, he had the manner of a trained scientist, the skills of a novice pilot, and the modesty of a lifelong also-ran. When he was invited to join the project, he told friends, "I guess I just stumbled into my big talent that I never knew I had." After the VO_2 Max test, it looked like Glenn wouldn't have to settle for local-hero status. He could be Daedalus. For this, he thought, if only for a year, I can chance dropping out of medical school.

On the day of the four-hour test, Glenn walked into Ethan's lab thinking he'd be lucky to finish three hours. Even training on his bike, he'd never been able to push past three and a half hours. Seeming oddly embarrassed, he stripped down to a pair of blue nylon shorts. Ethan noticed his nervousness, and as he rubbed alcohol over Glenn's chest to apply the EKG electrodes, he tried to reassure him.

"C'mon, Glenn," he said, "I can just see the glycogen jumping out of your legs now."

"Well," Glenn said, "I guess I'm as ready as I'll ever be." He dabbed Vaseline on his lips to keep them from chapping. "This is going to be hard enough. I don't want to start off with any disadvantages."

Ethan tried to distract him by doing calculations for the test. As lab assistants continued prepping Glenn—drawing blood, taking oxygen samples, weighing him on a scale—Ethan held a calculator and thought aloud. "Okay, Glenn, if we need three point three watts per kilogram and you're weighing in at one hundred fifty pounds, that means your oxygen intake will have to be. . . ." Glenn did the mathematics in his mind, and gave quick and accurate responses.

He clamped his riding shoes into the ergometer and spun the pedals. A line of syringes lay on a table nearby to take blood samples over the next four hours. In every way, the scene resembled the standard tests Glenn had observed during the summer. There was only one small difference. At the last minute, Ethan produced a canister of orange liquid and set it on a table in front of his chest.

"We're planning an experiment," Ethan said.

The drink was either a glucose-polymer concoction he'd devised—something like Gatorade—that would help Glenn extend his endurance limits, or it was a placebo—colored water—which would have no effect on his physical abilities. By taking blood samples, Ethan would be able to investigate whether a glucose-polymer drink would be of any help during the actual four-hour Daedalus flight in Greece. All summer long he'd been experimenting with the fluid using other pilot candidates who had reached the four-hour tests. He motioned to the 250-milliliter canister. "You'll be fine," he said. "The only thing I don't guarantee is the taste of it. That's not my department."

Glenn pedaled at 2.25 watts per kilogram to begin, hit the 70 rpm tempo of the metronome, and every fifteen minutes Ethan reminded him to down another 250 milliliters of the orange drink. Glenn placed a set of headphones over his ears and switched on a radio that he'd set up near his right hand. He closed his eyes, relaxed his arms and upper body, and let his mind go blank. His legs churned a perpetual motion.

After two hours, Ethan had tightened the strap on the ergometer so he could test his engine at 3.3 watts per kilogram, a slightly heavier load than anyone had ever attempted before. If the airplane was going to be overweight, Ethan needed to make sure the pilot could carry the extra pounds. He glanced at the chart recorders in another room. The blood samples looked good, the oxygen rate seemed normal, the heart rate averaged out at around 160 beats a minute, which wasn't bad. Glenn, it seemed, pedaled just below his anaerobic threshold. Ethan encouraged him to down another bottle of the orange drink and keep going.

An hour later, the arms on the chart recorders suddenly began to scribble intensely. The heart-rate monitor showed a steady increase in exertion. Dark rings had formed under Glenn's eyes and his body slouched slightly in the nylon webbing of the seat. Ethan heard Glenn's legs still cycling in cadence, but he also saw a large puddle of water—sweat—pooled around the legs of the ergometer. Glenn complained that his stomach ached. His thighs cramped, his calves burned.

"Okay, Glenn, give it a go," Ethan said. "Your heart rate's creeping up a little bit, but I want to keep going."

Glenn hung his head and squeezed his eyes shut.

"Okay, let's take this in segments," Ethan said. "You've just completed twelve minutes of the last fifteen-minute segment. Let's just finish the next three minutes and then go for another, okay?"

"I'm getting chills up my neck," Glenn said. "It's impossible to get comfortable." His legs kept rotating, but his torso, once relaxed, now bobbed and jerked spastically.

Ethan reached around to the back of Glenn's neck and felt his pulse. "Okay, Glenn, now push real steady," he said. "Turn the music up and try to forget everything for a while. Okay?"

"I'm having a lot of trouble keeping up," Glenn said.

After three hours and eighteen minutes, the legs quit cranking. Glenn slumped over and held his head in his hands. Ethan stripped off the EKG electrodes and sent one of his assistants to get him a Coke. "You're on your own now," he said. "You can pedal as much or as little as you want." The muscles in Glenn's calves flinched and pulsated uncontrollably.

Ethan leaned close to Glenn and said, quietly, "It's hard to keep secrets from medical students. Tell me, what was in the drink?"

"If it was an imitation," Glenn said, "I couldn't tell. If that's what you mean."

"To tell you the truth," Ethan confessed, "you got the placebo. You were running completely on your own fuel. After three hours your blood glucose was down, and you hit the wall. I'm afraid you just ran out of muscle glycogen. That's not entirely bad, though. I don't want you to worry about it. I mean, we got you past the three-hour mark. And with a supplement I don't think there's any question that we could get you past four hours."

"Look at my legs," Glenn said, pointing to his twitching calves. "I'm not doing that."

"That's okay," Ethan said. "During the last twenty minutes, you were just going on guts."

Glenn joined the team, but not without ambivalence. No matter how Ethan tried to explain it, Glenn always felt like a fraud, as if he'd failed the four-hour test. As usual, Glenn suspected that he'd seen his limitations, and though he wanted to be the Daedalus pilot more than anything, he wondered if he would ever really be able to fly like a Greek hero.

In late November, John invited the entire team to dinner at his home. After dessert, everyone crowded into the basement. There was the usual engineering chatter and every once in a while a melodious whistle rose from the group as they howled over a stupid sight gag stolen from Monty Python's Flying Circus.

Lois, whom John always thought was a little mousy around the team, was particularly quiet and, as usual, awed by the tightly wired engineers. The only newcomer was Glenn Tremml, and though he was greeted cordially at supper, no one from the engineering group believed he could single-handedly make their plane fly. The deadline was just a month away, and John felt agitated. Finally, the moment had arrived for a showdown between the manager and his team.

"I've been asking you guys all along what kind of chances you think we've got," John said, "and all I hear is a lot of brave talk. Now, I want proof. What kind of data do we have?"

He went around the room. He pinned Juan, collared Mark, buttonholed Parky, quizzed Bussolari. Emily still hadn't flown more than three minutes, but they all had excuses. No problem, they said. Don't get excited.

Of course, they all were pleased that Glenn could pedal at 3.3 watts per kilogram, which was slightly higher than Lois's 3.0. But since the airplane hadn't flown for more than three minutes, they hadn't been able to assess the power requirements for an actual flight. They needed five minutes in a steady-state cruise to store the data off the pilot's heart-rate monitor and estimate the power requirements.

Parky and Mark defended the airplane. Over the last month, they'd decided to extend the wings another twelve feet to give Emily increased lift, thereby lowering the power requirements. Why twelve feet? John asked. Well, they could have settled for half that, but if they could stretch the wingspan to one hundred twelve feet, then they could claim that they'd built the world's largest composite airplane. Besides, Parky said, every other HPA we built had a wingspan whose total footage ended with the number two. (More whistling, additional laughter.)

All the tests they'd done essentially showed that the airfoils held up, the prop worked well, the gears were good, the pilots looked tip-top. Everything was fine and, with a little work, the plane would fly for hours.

"Bullshit!" John snorted. "Where's the proof?"

"There's nothing wrong with the airfoils," Mark said.

"I don't think so," John said. "There are days we can't even get that sucker off the ground."

Once again, the engineers defended themselves. They'd used a video camera to take pictures of the wings' movement during flight, and they'd done some tow tests, flying Emily without a prop to get some indication of the stability and glide performance. Mark even took the wings out to the runway one night, hooked them up on the back of a tow truck, and used a stethoscope to listen to the air for turbulence as one of the UROPs drove him up and down the landing strip. The airfoils seemed okay. Well, all right, you couldn't hear much on the wing because of the rumbling from the truck, but still . . . The only problem, Mark said, was that dew and frost, which accumulated on the wings in early morning flight ops, created such a serious disturbance on his finely tapered airfoil that the slight condensation kept the airplane from flying well. If we wipe the wings perfectly clean, he insisted, performance will improve markedly.

C'mon, John thought. Give me a break. How do we know it won't re-condense in flight?

They argued that Emily was still a new airplane, and like any new airplane it needed tinkering to determine its peculiarities in flight, to assess the unknowns, to experiment. The pilots needed some time in the cockpit, too, to get acquainted with the controls and overcome the nervousness of flying a strange machine. Think about how much power is wasted in terms of a new pilot's anxieties.

"So you're saying there's still a heck of a lot of stuff we don't know?" John said.

"Yeah," Parky said, "we have a little tweaking to do. Just give us some time and we'll take care of it."

John fumed. The engineers stared back, certain and implacable. In just a few weeks they would have to fly out to California with two engines whose performance hadn't been adequately tested, and a big beer company waiting for its $200,000 air show. And still Emily couldn't travel farther than a housefly.

14

They flew from Boston to Burbank. In mid-December the UROPs rebuilt an old trailer used by MIT's five-man bicycle team, and drove Emily, dismantled, across country. High in the desert they gathered inside a Space Shuttle hangar and went to work immediately to make the airplane more buoyant. The change in scenery, from the urban confines of Hanscom to the remote, baked mudflats and big sky horizons at Mojave, invigorated the team. The empty vistas, ghostly tumbleweeds, and arthritic Joshua trees signaled an end-of-the-line appointment with some stern fate. An icy desert wind blew over the largest natural landing bed on earth—water-sopped, sun-hardened Rogers Dry Lake. Cowboy pilots rolled fighter planes playfully across the sky. In the mornings, when Juan and Mark visited the cold hangar and stared across the lakebed, imagining their quiet flyer gliding, sonic booms burst overhead. The ground trembled.

Inside the hangar, the engineers hung a United States flag on the wall and, beside it, the red flag of MIT. Late at night, when they couldn't gain security clearance from NASA to work in the hangar, they'd spread pieces of new fourteen-foot wingtips across the living room and kitchen tables of a spare apartment. New Year's passed with only slight deference to the holiday. A three-week deadline locked them in.

In early January, Gup Youngren took time off from Lockheed and joined the builders for the first time in months. He drove John through Panamint Valley to survey the landscape. They stood on the edge of the desert, where sands whipped and swirled in the wind. Although the testing site at Rogers Lakebed offered a generous arena for Emily to fly a ten-mile triangular course, John wondered aloud whether a straight flight across the desert would make a more dramatic challenge of the Albatross's twenty-two-mile record.

At the end of their tour, though, when Gup stopped at a Seven-11 for gas, John glanced at a copy of the *Los Angeles Times* and punched his friend.

"Holy cow!"

"What's the problem?" Gup asked.

"Look at this!" John said. He reached into the newspaper box and grabbed the stack. Front page—four pictures above the fold. The *Times* had announced Emily's arrival. His engineers hadn't even finished building the airplane, but already their bluff had been called.

That afternoon, newspaper, radio, and television reporters from across the country phoned John's apartment all asking the same question—when would they fly? John immediately assigned one of the UROPs, Peggie Scott, who had just graduated from MIT, to the job of press liaison. Together, John and Peggie rehearsed responses. While the engineers puzzled over Emily's weight problems, Peggie scheduled a press conference and explained the project to bedazzled journalists. But there was still that one question she couldn't answer—when? Emily had never flown for more than three minutes. What could she say?

The January rains saved them. High winds and muddy conditions on the lakebed gave them a convenient alibi. It depends on the weather, she'd say. Sure enough, the West Coast papers labeled them whiz kids, too.

On a runway outside the hangar, Lois McCallin continued testing Emily. Despite the rains, they could risk trial flights across pavement. She wore a satiny electric blue hat with gold wings, and orange tights that glowed like hot coals in the cockpit. Flirtatious teasing, once her bond with the engineers, no longer sweetened their banter. Over the last six months Lois had earned respect from the team for her concentration and steadiness at the controls. Perhaps no other pilot with so little hope could have gained the support she had from the ordinarily demanding and officious crew. Perhaps they'd engaged in denial for so long about their own abilities that they couldn't see Lois's limitations either. She had never flown for more than three minutes, yet the engineers couldn't quite believe Lois might really fail.

Instead of criticizing her, the engineers blamed themselves for weight excesses and high power requirements. When it came time to improve the wings, they'd tell Lois they were doing it for her. When the engineers worked on the seat, they'd say they were doing it "for Lois."

Mark Drela watched her trial flights on Mojave's runway from the back of a truck. Gup grabbed the tail boom, and two UROPs stood beneath the wing tips, one hundred feet apart. Once Lois started to pedal and the great wash of wind blew from the propeller, Gup leaned into the boom and the wing runners, who held the tips aloft to keep them from bending down and breaking, started to jog with him. Emily rolled ten feet, twenty feet. Mark cocked an eyebrow. Lois's legs flashed inside the cockpit. The wing runners galloped. From behind, Gup suddenly heaved the airplane as if he were making a javelin toss and ducked to keep his head from smacking the rudder. She was up.

"Uh-oh, uh-oh," Mark moaned. The right wing drooped and the fuselage jostled under Lois's straining muscle. The airplane climbed about three feet and after a minute, drifted to the pavement. Mark leaped over the back gate of the truck and ran out to catch the right wing before it touched ground.

"Gotcha! Nice, Lois," he said. "Feels kinda doggy, huh?"

Steve Bussolari watched from a truck that rode behind the airplane. He'd worked with Lois for seven months, teaching her to fly gliders so she'd make a smooth transition to the cockpit of an HPA. As he talked to Lois over the radio during flight ops, he fed a consistent stream of support: "Looks good, Lois. Right rudder, give me a little right rudder. That's it. Good. Good, Lois. You're a little nose-high. . . . Now you've got it. Good, Lois." But after the second trip down the runway that day, another sixty-second struggle, Bussolari clipped the radio mike to the dashboard and admitted to himself that they had a real problem. "Okay," he muttered, "we got power trouble." He stepped out of the truck and slammed the door. Fat Emily drained their engine.

Mark and Ethan Nadel paced around the wing. Mark brushed his hand over the top surface of his airfoil and shook his head. The wing was scratchy with frost.

"That's at least a thirty percent drag influence," he whispered.

John Langford in 1970 (age thirteen) at a model rocket contest in Atlanta, Georgia. (*Langford family collection*)

Guppy and John, 1976. John, now a freshman at MIT, is learning how to fly radio-controlled model airplanes from Harold ("Guppy") Youngren. (*Langford family collection*)

MIT Rocket Society meeting, circa 1975. John is at lower left, Guppy is center left. (*MIT Rocket Society Archives*)

John Langford with his Athena H scale model at the 1978 World Championships in Yambol, Bulgaria. (*Trip Barber*)

The rocketmen's first experience with human-powered aircraft. Attempting to fly MIT's BURD in the fall of 1978, they attach model airplane engines and put Guppy in the cockpit. Bob Parks is at left, Guppy at right. (*John Langford*)

The end of the BURD, December 1978. Bob Parks compresses the BURD wreckage.
(*Bob Parks personal files*)

Chrysalis at rest in its hangar, with its 1/8 scale development model, June 1979. (*Steve Finberg*)

Chrysalis soars in a flight test, June 1979. (*Steve Finberg*)

Mark Drela designed and built the propellers for the 1983 Monarch, and here wears them as rabbit ears. (*Allan R. Shaw*)

The Daedalus team, summer 1985, at an early meeting of the working group. *Standing:* Bob Parks, Jim Wilkerson, Jim Hilbing, Harold ("Guppy") Youngren, Bryan Sullivan, John Tylko, Steve Finberg. *Kneeling:* Mark Drela, Juan Cruz, Debbie Douglas, John Langford, Steve Bussolari. (*Steve Finberg*)

Juan Cruz with the experimental spar section that features three tubes kissing, March 1986. (*Steve Finberg*)

Lois McCallin piloted the Michelob Light Eagle to three world records on January 22, 1987. (*Michael Smith;* inset, *John Langford*)

Guppy, Mark Drela, Juan Cruz, and John Langford after the Michelob Light Eagle completed successful test flights and set four new world records at Edwards Air Force Base in California, January 1987. (*NASA*)

The Michelob Light Eagle in the Smithsonian's National Air and Space Museum, August 1987. (*National Air and Space Museum, Mark Avino*)

The Light Eagle approaches the beach after its overwater test flight off Rhode Island, August 1987. (*Barbara J. Langford*)

Dari Shalon emerged as a leader in the assembly of the two Daedalus aircraft during the summer of 1987. Here Dari assembles one of the Daedalus horizontal stabilizers. (*Donna Coveney, MIT News Office*)

By the time the first Daedalus airframe was rolled out in October 1987, the Daedalus team had grown to twenty-eight.
Back row, L-R: Tom Clancy, Siegfried Zerweckh, Bryan Sullivan, Peter Neirinckx, Guppy, Tom Schmitter, Peggie Scott, Steve Finberg, Tim Townsend
Middle row: John Langford, Jamie Pavlou, Mark Drela, Juan Cruz, Steve Bussolari, Matt Thompson, Steve Darr, Mary Chiochios, Dari Shalon, Marc Shafer, Ethan Nadel, Jean Cote, and Jack Kerrebrock
Front row: Ellis Langford, Lois McCallin, Glenn Tremml, Erik Schmidt, Greg Zack, Kanellos Kanellopoulos (*MIT Lincoln Laboratory*)

Grant Schaffner and Dari Shalon assemble tail surfaces of Daedalus. (*Donna Coveney, MIT News Office*)

Mark Drela and Jean Cote install controls in the Daedalus airframe. (*Donna Coveney, MIT News Office*)

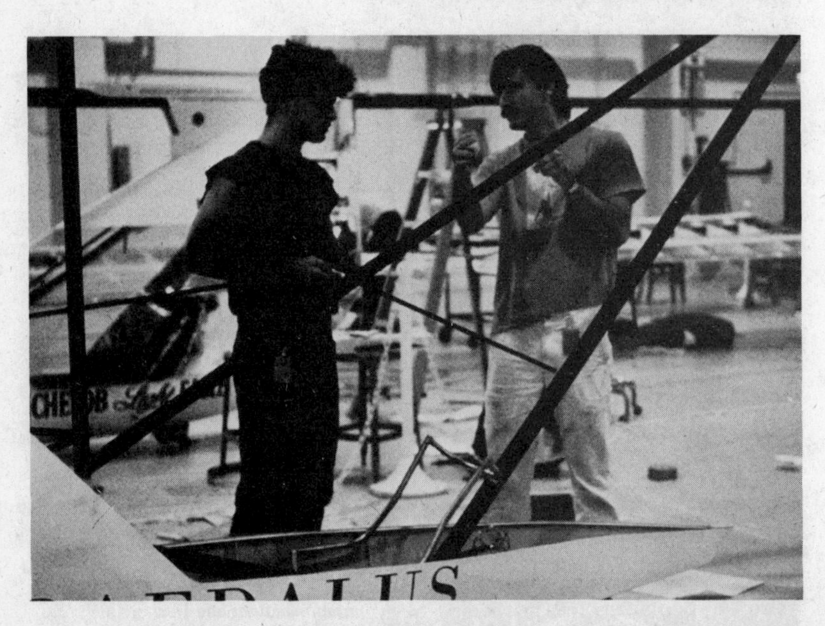

Tom Clancy (*left*) and Guppy confer during construction of Daedalus 88. (*Chuck O'Rear*)

"No way," Ethan said.

Mark looked at the wings. His stomach gurgled.

"Lois is basically dying up there."

The team turned the airplane around for another flight. Parky sat in the back of a van to keep warm, making subtle observations about Emily's flight characteristics. He looked like a solemn guru. Mark caught Lois's eye and motioned toward the van. "That's the tour bus," he said. "First class. I think they're serving drinks, too."

Lois clipped her cycling shoes into the pedals and waited for instructions.

"What is this, a sightseeing vehicle?" Mark teased. "Get up there and work."

In a fuselage shaped like the body of a goldfish, behind a transparent screen of 0.5-mil Mylar, Lois looked content. She leaned back on a blue cushion, cut precisely to fit her torso. Outside, a team of engineers dressed in wool caps, sweatshirts, and gloves huddled under a wing. Clouds of cold vapors rose from their lips.

Mark sent one of the UROPs for a bucket of hot water and a chamois. He touched the top surface of the wing and flicked off another flaky white layer of ice.

If they could decrease the power requirements by just 7 to 10 percent, Lois would pedal beneath her anaerobic threshold; flight times could easily quadruple. As Bussolari explained it, at a certain point the muscles in Lois's legs built up so much lactic acid that they ceased to contract. Decrease the weight a mite, he told them. Reduce the drag. "A pilot can't overcome that barrier," Bussolari told Mark. "No amount of willpower will overcome it."

Lois's brother had driven in from L.A. to watch the runway trials. Her sister flew in one day from Philadelphia. Then her father arrived from Milwaukee. Another sister came in from San Francisco. The McCallin family, an entire clan of triathletes and runners, had always supported Lois's efforts in athletics; she was the last one in the family to take a serious interest in fitness. But with the advent of Daedalus, they prepared for celebration. In early January, the McCallin name was spoken with familiarity around Edwards Air Force Base; photographs of Lois with the team were on display in the gift shop. Her family watched the trials, like the rest of the team, and waited for engineering magic.

Then one calm dry day, the team carted Emily out to the lakebed unannounced and told Lois to fly as far as she could. The McCallins slipped past the security guards. Lois flew for eight minutes—an all-out gut-busting sprint—and the team went crazy. Lois's father jumped on the recumbent bike and rode it around the hangar. John insisted that they could start doubling their times with every flight. For once, it appeared that Lois might challenge the record.

"I have to do it for them," Lois said. "I feel like it's my responsibility to make this airplane work. The engineers can do it all, as far as building the parts or improving the wings. They're wizards with that stuff. But they can't fly it. That part's up to me."

On January 8, the logs showed that Lois's longest flight had been eight minutes. That same day, at the University of Connecticut Medical School, Glenn Tremml convinced one of his professors to let him take his cardiovascular exam a couple of hours early. "I have to learn how to fly this new airplane," he explained.

Glenn rushed to catch a flight leaving Hartford for Burbank. While he dozed on the plane, a news summary flashed on a movie screen. A stranger across the aisle noticed Glenn's gray Michelob Light jacket, bearing the Daedalus logo, and tossed an empty beer can in his lap to jostle him awake. The guy was pointing at the screen. The news summary. It was Lois on the runway, pedaling Emily. Everyone was smiling for the cameras.

When he reached Burbank, Glenn took the Antelope Valley Shuttle to the Desert Inn in Lancaster, a couple of miles away from the Sierra Pines apartments where the team was living. He unloaded his bike from the van, slipped on his backpack, and pedaled off in the dark. It was midnight. Lois, Bussolari, and John and his family were already asleep. The engineers, who'd transformed their apartment into a shop, invited him in, where they labored intently over the new set of wing tips. He bedded down amid assorted tools and sheets of Mylar. Four hours later he was awakened and hustled out to a dark, frosty runway.

In the glimmering sunrise, Glenn rubbed the sleep out of his eyes. He stripped off his sweats and grabbed onto Bussolari's hands. The young professor gently guided him through the small rectangular window that opened into the cockpit. "Squat. Okay, butt first. Hold on tight. Head down. Slowly. Now, lean back." The airplane was so fragile that Glenn couldn't reach into the fuselage for support. He eased nervously through the opening, leaned onto the seat, and painstakingly lifted his legs in, setting one foot on a pedal, then the other. He barely fit.

Glenn's shoulders touched both walls of the Mylar windows. When he lowered his hands to feel the controls by his side, his knuckles brushed against Parky's hard, orange Kevlar tub. One of Juan's black carbon tubes jutted up diagonally from between his legs. Flickering twelve inches in front of his eyes were Finberg's flight instruments, tiny liquid-crystal display monitors. Half the pod had been covered with aluminized Mylar to discourage sunlight. He felt like he was balled up in an oversized egg. He sat less than an inch off the ground. His legs, hands, and head were free to move, but nothing else. His heart jumped inside his chest.

Electronic sensors would gauge his speed and distance off the ground. Bussolari handed him a heart-rate monitor to clip around his chest to measure power. By looking at the readouts while he was in the air, Glenn could cross-reference the numbers and estimate the most economical airspeed—theoretically, whenever his heart rate hit 160 beats per minute, he should have Emily flying about ten knots.

Glenn grasped the control stick, about the size of an index finger, and jostled it side to side. The rudder wiggled.

"Does this button do anything?" he asked, thumbing a black tab at the end of the stick.

"That's push-to-talk on the radio," Bussolari said. "We'll deal with that later. Just remember, it's like we talked about before. Right hand on the controls. Your left hand is free. Shoot for ten knots speed. Watch the displays."

"It's gonna be easier than I thought," Glenn said.

The engineers grimaced. They explained that any crash was likely to be fatal: they'd kill him. Then they rolled him out to the takeoff strip.

Glenn flew five times. Every takeoff was a heart-wrenching experience. The wheels scraped along the pavement, and the cockpit rumbled like a drum. Glenn pedaled rapidly to gain power for liftoff and his heart rate jumped to 180. But once he was up, the wheels quit spinning and he was enveloped in a kind of surreal stillness. Emily floated. As he pedaled and his heartbeat slowed, he glanced out at the wings and saw them bend. The tips disappeared and reappeared. Way, way out, the wings flexed at his command—right rudder, left rudder, right aileron. So thin, and so long, they looked like fingers trembling, flicking at the clouds. It was spectacular. Alive. My God! he thought, I'm in the air.

Glenn flew four more times that day. The next day he made six quick flights. They couldn't wait to test him on the lake bed. John and Brian Duff drove to a press conference in Los Angeles and announced that Emily would attempt a record flight within ten days. If it didn't rain, of course. If the winds were right. The next day, Glenn flew Emily over the lake bed—five miles in eighteen minutes.

The engineers had succeeded in reducing Emily's weight by only two pounds, yet flight times were now doubled. For Glenn, at least, the airplane's power requirements had dropped below the anaerobic threshold. On January 18, he flew eighteen miles in forty-five minutes, the third-longest human-powered flight in history.

The engineers were exhilarated. They had played all their cards. The mast, a long stem strung with wires that held the wings off the ground, had come off: a major source of drag, eliminated. They'd built a more streamlined fairing on the fuselage. They'd become obsessive about cleaning ice and water off the wings with hot-air blow dryers. Mark and Juan finally completed a set of longer wing tips to give the airplane greater lift. And, of course, they had a stronger engine.

That night a group of team members confronted John. If official NASA record-keepers had been on the lakebed that morning, Glenn could have flown five more miles and captured the record for absolute distance. Why did we have to stop him? they asked. Why do we have to wait for journalists and camera crews to show up?

"This isn't a show-and-tell game for the press," one of the engineers argued.

"Yeah, but we've got to deliver for the sponsors," John said. "If we notify the press, I'll bet we get on the network news."

"We're not out here to be on the evening news," the engineer snapped.

The next day, Brian Duff alerted the media. One of their pilots would attempt the record flight on January 21. No, he couldn't say which one.

* * *

Although they shared the same apartment with Bussolari, Lois and Glenn didn't have much contact with each other. They trained separately, cycling and running on different days. Glenn flew the flight simulator regularly, out of necessity, improving his flying technique on an earthbound ergometer. Late at night in the hangar, he could be found pedaling furiously, tilting the joy sticks on the simulator, watching a blue video monitor. A cartoon Mojave slipped beneath his cartoon airplane. Crashes were okay there.

As Glenn practiced, Lois burrowed even more intently into her training regimen. At dinnertime, Glenn thought Lois seemed awfully reserved. If she fretted about losing the cockpit to him, she rarely said so. But she did have mixed feelings about his sudden arrival on the team. "Glenn's the guy with the legs," she'd allow. She knew she was no competition. Only the engineers could make her a contender for the record.

Naturally, the 21st was Glenn's day to fly. He and Lois had entered a rotation, staggering training and flight-readiness days. But Glenn was the chosen one. Lois, if she flew at all, would fly second.

During the early morning of the 21st, one of the team members walked out to the desert with a Mylar streamer and checked the winds. The streamer blew almost horizontally; weather reports from NASA predicted winds at over five knots, the cut-off for Emily's fragile flights. John made the decision to pass Glenn's day, and notified Lois that she'd challenge the record on the 22nd. Hope against hope, Lois thought.

Loaded up on a strict diet of rice, bread, potatoes, apple juice, and seltzer, Lois was well prepared to confront MacCready's twenty-two-mile record. She wanted it. She couldn't disappoint the team.

The next morning, before the sun rose, she warmed up behind the mountainous fuselage of a 747. From atop the Shuttle's hangar, members of the team watched Lois pedal a few strokes in the cockpit; Emily looked like a spindly prehistoric insect twitching its antennae at the rear end of a dinosaur. Three minutes before takeoff, UROPs and engineers hovered over the wings with hot blow dryers, shrinking the skin and swiping off beads of ice and water. They wanted Emily to be perfect for Lois. It had to be.

She ran back into the hangar to finish stretching, and John warned her to keep still until he could redirect journalists who were scouring the grounds for pictures and comments from the female pilot.

"This is it," she said. She quietly steeled herself for the flight. She was trembling slightly as she boarded a van and was driven to a concrete ramp on the edge of the desert. The ramp led down through a set of wiry gates and ended abruptly on the crusted red surface of Rogers Lake. She climbed into the cockpit and shut out the noise of the engineers' chatter. She tested the pedals.

"Okay," she told Bussolari over the headset, "I'm ready to accelerate."

The long black prop spun. Juan lifted up one wing; another teammate, Jean Cote, held up the other. Photographers standing atop a flatbed truck squeezed shoulder to shoulder, snapping pictures. Emily started to roll. To set an official record, Lois couldn't have a boost from the tail boom. It was the first time she'd ever attempted takeoff under her own power. The cockpit filled with a stormy rattling as the wheels scraped concrete. Jean and Juan jogged alongside, hands bolstering wings. At twelve knots, Lois passed the gate, leaning way back in the seat, churning as hard as she could. The wings lightened in Juan's hands and Emily started to climb.

Lois rose ten feet and felt Emily float. She heard the prop sing, a sweet whistle. She looked at the mountains in the distance, the flat red ground beneath, and a promising gift of cold blue sky.

"See the mountain?" Bussolari radioed. "Okay, give me about twenty degrees of that. We need a little right turn . . . Give me a little more right turn."

The airplane felt stable. She brought Emily through the first three-mile leg without any trouble. She'd never flown so far. Lois took the airplane to twenty feet, twice as high as she'd ever flown it. She rounded a corner. In the race to the second corner of the triangular path the airplane hit a gentle thermal and she climbed to thirty feet. Twenty minutes passed, but time seemed suspended in the cockpit. The windshield fogged up and her heart rate jumped from 167 to 178 beats per minute. Bussolari's directions came in quick, steady pulses. As she rounded the second corner and brought Emily through the final leg of the course, Bussolari called.

"Can you make it around again?"

There was no answer.

"Okay, Lois, we need to make a decision now about whether we're going to make that last turn. Whatever you feel is safest."

She was winded, awash in sweat. Driving into the third leg, she'd felt confident that she could muscle past the anaerobic threshold—the engineers had stripped two pounds off the plane, they'd added longer wing tips. Couldn't she at least make it twice around the course? Lois looked for the third corner. Another minute or two might be possible. But another turn would sap her. Did she have to stop where she began or could she fly further? Could she fly the circuit again? She looked at the heart-rate monitor. Her legs felt full and hot. Her lungs couldn't hold more air.

"I think we should consider landing," Bussolari called. "We're probably going to guide you upwind at the last turn point. That's thirty-four minutes, Lois!"

She closed in on the marker, and eased Emily down. The wing runners, who were appointed to catch the wings to protect the spars from snapping, had to sprint to catch up with her, as she kept cranking and scooted quickly over the lakebed. Lois had flown 9.9 miles in 37 minutes, 38 seconds. Team members piled out of their vehicles and ran excitedly toward the cockpit.

John held off the swarm of journalists, saying that Lois had just made "a heroic effort." He announced that she'd set three world records for human-powered flight: a closed-course record of ten miles; straight-line flight for a woman; and duration for a woman. He wasn't certain whether the National Aeronautics Association would accept the category records—he'd just sort of made them up. But the press was ecstatic. Photographs of Lois went out across the wire services. *Glamour* magazine bought color pictures from a freelancer. So did *Sports and Fitness*. Editors at *The Boston Globe* prepared a layout to run her picture on page one.

"It's too much," Lois said afterward. "*Glamour* magazine? That's ridiculous."

There was a party that night and the engineers toasted their champion with champagne. But even as people began to talk about the flight to Greece, Lois knew her place on the team was no longer secure. Despite the celebration, Lois was no longer the project's "super athlete." In truth, although nobody said so, she was finished as a pilot on the Daedalus team. The engineers had failed her.

Glenn awoke with a slight buzz from the brownies he'd been eating. In contrast to Lois, he didn't take his nutritional needs too seriously. He also had a sweet tooth. An all-spaghetti-and-brownies diet sustained him, happily, for the days at Mojave.

One of the engineers greeted him at dawn with news that a 747 had tested its engines unexpectedly the afternoon before on the lakebed. The blast had covered the runway with rocks and brown dirt, making a smooth take-off impossible. But they'd planned a strategy to distract the press for an hour while the team cleaned the pavement. He'd be driven to the hangar, like an astronaut, and taken into a locker room at NASA headquarters to make it appear that serious preparations were underway. All he had to do was sit and wait.

Okay, Glenn said. What would he say to the press anyway?

At 8:00 a.m., he left the locker room. As he walked with Bussolari through the hangar, a voice sounded over NASA's loudspeakers: "Attention! This morning Glenn Tremml will be flying on Rogers Dry Lake in an attempt to break the seven-year-old world record for human-powered flight!" Glenn thought the voice shuttered eerily over the desert.

"Gotta do it now," he said.

They put him in the back of a van and drove him slowly the one hundred yards to the airplane. His breathing was shallow. He quickly downed a pint of water from a plastic jug and peered out the window. A half-dozen engineers walked around the wings, trailing a noisy generator that energized blow dryers. Gup rattled through a toolbox for a set of pliers. Juan tugged at a generator coil. A group of UROPs with brooms swept the concrete.

Glenn tiptoed barefooted across the frozen pavement. He took off his sunglasses, then his sweats. Bussolari helped him into the cockpit, and watched him clip his flying shoes onto the pedals. As the engineers unfastened the mast and lift wires, Glenn paused to examine the hydration system filled with five pounds of water above his head. He arranged the long plastic tube from which he'd drink so the tip dangled in front of his face; he could easily lean forward and take a gulp while he flew. He pursed his lips to the nozzle and sipped.

He blew warm air into his hands and began to pedal slowly. Juan finished heat-shrinking the wings and pulled a chair away from the plane. Emily's skin was so delicate that any change in temperature caused the plastic to wrinkle; Juan made a final inspection to make sure everything was tight. A plastic sheet of Mylar was taped over the opening in the cockpit, and someone turned a blow dryer to those wrinkled edges. John walked briskly toward a van with a wristwatch clenched to a clipboard. Glenn gripped the controls. He felt a cold line of sweat trickle down his arm.

The team cheered when Glenn lifted off the runway and climbed over the lakebed. "You are now on course," said a voice over his headset. One of the official observers, an Air Force captain, squeezed a button on his watch.

John, Bussolari, Parky, Juan, Jean, and a camera crew piled into the chase van to follow. John wore a clean white shirt and blue tie with tiny designs of the Monarch HPA on it. His blond hair was neatly combed. Everyone else wore the rumpled clothes they'd slept in. Juan pulled the hood up on his red sweatshirt. Parky tugged a wool cap over his head and shuffled into a corner. Their faces were drawn, and they had circles under their eyes.

Glenn glanced at his monitors. Speed, fifteen knots. Altitude, ten feet. Heart rate, 171. He knew his lactate threshold hovered around 172. The airplane demanded too much power. Maybe the wings had collected dirt on takeoff. After fifteen minutes, he pressed the button on the control stick.

"It feels too hard," he said.

There was no answer. The chase van pulled up suddenly beneath him.

"Jiggle the controls," Bussolari radioed. "The alignment's off a little on the wing tips. Maybe they'll snap into place."

Glenn tugged a control by his left hand—the ailerons, delicate control surfaces built into the end of the wings, jostled suddenly. The pedals lightened, his heart rate dropped.

Ten minutes later, he heard his right shoe scrape against the tub. His foot had slipped. The pedal itself was just a bar that locked onto a cleat on the bottom of his shoe, but it was a new device, and somehow never tested. Glenn yanked his foot off the pedal and tried to reattach the shoe as he kept cycling with his left leg. He radioed the van for help, but the transmission was cloudy. The guys in the van watched Emily descend rapidly from twenty feet. Glenn pumped hard with his left foot to bring the airplane up, and while he pedaled he reached down to reattach the right shoe. The airplane tilted on a quick glide path toward the lakebed.

The altimeter measured his descent, six inches a second.

"You're going to touch!" Bussolari shouted over the headset. "You're going to touch! . . . Six inches!"

Just as Emily's nose dipped to the height of a playing card above the dirt, the cleat snapped into place. Glenn pumped with both feet and tussled with the controls to pull the nose up. It was all he could do to squeeze his feet inward on every revolution and keep the shoe attached to the pedal. He'd only covered six miles.

His foot kept creeping off the rod, but from the ground nobody noticed. Emily again looked beautiful, cruising the course, taking long, slow turns around the markers. The engineers were mesmerized.

Bussolari and Parky realized that something had gone wrong with the pedals, but they couldn't help Glenn from below. He learned to control the problem by climbing an extra three feet after every slip and making corrections as the plane descended. Just when his right foot reached the tip of the rod, he pedaled twice with his left foot, yanked the right one loose, reached down, and slipped the cleat back on again.

From the chase van the crew noticed the fuselage wobble, but nothing more. Glenn had quit talking to save his strength, and the engineers decided that whatever problem he'd had was finally solved.

Forty minutes into the flight, Glenn leaned forward to take his first drink of water. He couldn't draw more than a sip. He pulled the hose into his mouth and left it there. He sucked and breathed deep draughts of air through his nose. Nothing. He tilted his head back, hoping to unstop the tube. He squeezed the valve and sucked. The stop mechanism in the hose, rigged to ensure that the liquid didn't leak into the cockpit, apparently worked too well. Great, he thought. Five pounds of useless water and a busted shoe. The windshield had fogged up and he couldn't see, so he simply closed his eyes and listened to Bussolari's instructions to guide him. He and Emily flew on, hobbled, blind.

The crew in the chase van, unaware of how serious Glenn's problems were in the cockpit, tried to guess when he would pass the twenty-two-mile mark. They'd lost track.

"I think the coast of France is in sight," said Jean Cote, referring to the Albatross's flight across the English Channel.

"I'd say we're not even halfway," said Parky, who was thinking about the sixty-eight mile flight out of Crete.

The van fell silent as the crew followed Emily's progress, mile after mile. They'd become spectators. When an hour and a half had passed, Parky spoke up.

"Tell him he's hired, John."

Bussolari put his feet up on the dashboard and leaned back in the front seat. "I'm going on strike now," he said. "I want more pay."

It was most uncomfortable for engineers to be idle. In the absence of scientific analysis, the flight had turned into an athletic event. After thirty miles—Glenn's third lap around the course—they were plainly bored. Flying wasn't as interesting as engineering. They were excited only when the airplane, once again, dipped suddenly to within three feet of the dirt. When it didn't rise, Parky leaned forward: "What's wrong?"

"Bear down now, Glenn," Bussolari said.

In the airplane, Glenn panicked. His foot had slipped again. He realized he would have to climb back up to eight feet, level off, and try to snap the shoe back into place. But it would take an enormous surge. He cranked Emily up to six feet, tried to establish a rhythm, and prepared to let his foot fall free, to reach down and reattach it. But he was afraid he'd miss. His muscles had cramped, and he heard unsettling news over the headphones.

"We've got a massive down draft," Bussolari told John, pointing to heavy silver-blue clouds moving slowly down from the mountains. "It's gonna start raining." The crew didn't realize how desperate the pilot had become.

"I think I need a radical pedal change," Glenn gasped. In the van they heard just another garbled message over the radio.

"What did he say?" Jean asked. "What's a radical pedal change?"

Glenn yanked his foot off the rod, but his legs suddenly felt lame and rubbery.

"You're losing altitude!" Bussolari called. The crew heard him kicking inside the cockpit, and Emily drifted down, wobbling, veering off course. The moment Emily's wheels scratched the lakebed, the wing runners hurtled out of the van and gave chase.

An Air Force captain who'd served as the official record keeper radioed the van, and John wrote these numbers on his clipboard: 37.2 miles, 2 hours, 13 minutes, 14 seconds. They'd obliterated the Albatross record. They'd beaten MacCready by fifteen miles.

Inside the stilled cockpit, Glenn lowered his head, feeling defeated by the pedals. He could have gone farther. But as a crowd gathered around Emily's wings, he stripped off the Mylar window and eased his head into the cold air. A hoard of photographers and reporters hustled across the empty lakebed.

Glenn met them, half naked, on the crusty flats, drawing them away from the airplane. He'd lost four pounds during the flight, but he didn't hesitate to step into a circle of gaping photographers and conduct his first press conference.

"I certainly thought my effort was great," he said, "but it wouldn't have been possible without these engineers here. You wouldn't believe it, but just a few weeks ago our maximum effort was a three minute flight . . . I guess that's why they call it an experimental aircraft."

The press pulled in, cameras tilted down at his legs, questions fired from the group. Hey, Glenn, how'd you do it? Glenn, did you say something went wrong with the pedals? What's next? Tell us, how does it feel to fly?

"How does it feel?" Glenn repeated. He smiled. Now that was the real question: Snap! Click! Whir! Scribble! "It feels fantastic."

The flight of Emily made headlines from Los Angeles to Washington. It made the major news magazines. It made network news. Glenn Tremml, partly amazed, partly amused, was the day's American hero.

On their way back to Boston, Bussolari broke the news to Lois: she couldn't continue as a Daedalus pilot. They'd take Glenn, but they really needed athletes who were even more powerful than he.

"I understand," Lois said. "You don't have to explain." She had already planned to go back to work at the investment company.

When she returned to Fidelity the next week the corporate environment felt alien. Her co-workers seemed dazed by their own projects and late night deadlines. The project she'd just left had been so different—a privileged life, somehow.

A few friends in the office taped news clippings of the record flights on the walls. But for the most part, her colleagues were so distracted by their own work they hardly seemed to notice what she'd accomplished in California. Within a few weeks, she joined their routines, and longed for Daedalus.

Glenn had expected a media tour. Not long after the flight, Bryan Allen, the pilot of the Albatross, dropped by the NASA hangar and told stories about the cross-country travels DuPont had arranged for him. Bryan enjoyed a few weeks of celebrity, going from town to town, giving interviews across America about the English Channel flight. Glenn couldn't wait to see what Anheuser-Busch had planned for him.

But within a week of the flight the beer company had lost interest. In fact, the sponsor not only discarded any plans for a publicity tour, it also wanted the Michelob van loaned to the project returned to St. Louis immediately. There'd be no press kit, no clipping service, no tour. The marketing managers at Busch saw little to gain from involving the company in the next phase of the Daedalus Project.

Undaunted, Glenn volunteered to drive the van to St. Louis anyway. Maybe there'd still be an opportunity to do an interview or two for the newspapers there.

He drove for three days. In Edmonds, Oklahoma, famous for a recent post office massacre, he stopped overnight to stay with his girlfriend's sister and brother-in-law. He wore his Daedalus gear—black tights and sunglasses—to entertain the family's kids, and he called the local newspaper and radio station to say he was in town. The kids enjoyed the show; the editors said, sorry, that's old news.

When he arrived in St. Louis, nobody at Anheuser-Busch was even waiting to shake his hand. Glenn left the van at Busch headquarters and made his own reservations for a flight back to Connecticut. In the weeks that followed, when he recounted his adventure to friends, he always ended with the same self-effacing tale about the trip to St. Louis. "It was all very impressive," he would say. "I'd never been to California before."

The engineers stayed an extra week on the lakebed, stuffing Emily with SONY 8mm camcorders, remote-control gadgets, experimental fiber-optic wires, and data collection systems. NASA politely urged them to leave. At last, they had all the new data they needed to redesign the wings and spars. If Emily hadn't been the perfect airplane, Daedalus might be.

Unfortunately, costs for the final phase of the Daedalus Project—construction of two new airplanes, paying the salaries of an expanded crew, hiring athletes for the pilot candidate team, and sending the entire team to Greece for up to two months—would total $900,000.

MIT and the Smithsonian delivered $60,000 to tide the team over in return for a series of technical papers about the airplane. But Emily languished in the NASA hangar in California for three months because nobody had the cash to haul her home. In April, the National Air and Space Museum arranged to place Emily in an exhibition, and the airplane was trundled to the East Coast in the back of a truck.

As Emily was mounted at the Smithsonian, sharing air space with the Wright Flyer and the *Spirit of St. Louis*, John helped organize public relations events to keep the project's name in the news. Glenn was interviewed for a story in *The Washington Post*'s Style section. The same day he visited the Air and Space Museum, and spent a few hours standing beside a public video monitor set up next to the airplane; not one person, he confessed, recognized him. Juan was emotionally overwhelmed seeing Emily next to the Lindbergh airplane, though, and he spent hours just sitting in the museum, savoring his accomplishment. The flight had been little more than a passing entertainment in America, one day's amusement for the millions.

Over the next few weeks, John outlined the schedule for construction, test flights, and Greek operations of the Daedalus airplane, and, once again, faced the enigma of their existence—under the threat of bankruptcy, his team had no product to sell potential sponsors, no services, no guarantees. For $900,000 Daedalus offered only an inspiration, a risk, and a dream.

There were no buyers.

15

Over the years, Hanscom Field had become an ageless place for the team's original engineers, like an old ballpark that's open for only a season but amasses more memories than a heart can hold. Nine years had passed since Gup Youngren climbed into BURD and felt the canard snap as he pedaled down the runway behind a pickup truck. Eight years before, John had invited MIT's aeronautics faculty out to watch the first flight of Chrysalis. Four years ago Juan sat in the hangar and cradled his head in his hands after the first Monarch airplane crashed, time and time, on a blank lick of pavement.

In 1987, summertime at Hanscom again brought many of the same people out to the field, though probably for the last time. Gup left his home in California to take an apartment in Watertown, Massachusetts, and replaced Parky as the director of Daedalus' engineering team. Parky had married one of Gup's former college girlfriends, and sent his regrets; he couldn't sustain a marriage in California and work at Hanscom for another summer.

John, who now had a two-year-old son and a Ph.D., left his family in Washington. Mark, whom everyone teased unceasingly about his receding hairline, took time away from the Institute to shape balsa on a lathe at the Hanscom shop. Juan cut his shoulder-length hair, moved into Gup's apartment, and wondered about what he'd do when the project ended. He had ruled out working for defense-related companies, and he knew he might have no other choice than to live in Europe. Even without guaranteed funding, the Daedalus Project allowed a number of them to delay or ignore professional decisions for at least another year.

The team's roster grew. Several of the undergraduates and volunteers rented a house in Lexington and settled into a communal existence that fit their odd hours and esoteric, almost otherworldly interest. Sequestered in a wooded suburban neighborhood, the project house quickly filled with the belongings of managers, engineers, and undergraduates. The curtainless rooms with their three-legged chairs and clocks that had quit working filled up with the accoutrements of youth: crystals and compact disks, calculators and Swatch watches, books of Greek language, Grateful Dead songbooks and guitars, model airplanes and holograms, soap bubbles and piles of dirty clothes. Only on one desk—John's—was there evidence of creeping adulthood: stacks of newspapers, *The Economist*, *The New Republic*, management tips from business publications about executive burnout.

Summertime at Hanscom had always been a time for confronting rare possibilities and seizing dreams. It was a time for vision. The original engineers were growing older, but at least for one more year it seemed that the flying season was full of promise, a time to test new wings.

One afternoon, in 90-degree heat on a cleared pocket of land off Highway 128's High Tech Corridor, Gup Youngren cleaned shop. He rambled out into the solar home, where UROPs cut foam and shaped ribs, scattering white beadboard in a red-faced flurry.

"It's really hunky!" he shouted. A half-dozen young builders stopped their work around the shop and stared.

Styrofoam skittered along the pavement and danced in the wind. "Hunky" hardly seemed to describe material that the breeze batted around like waste paper. But inside the solar home, where the construction of a new airplane hit a quick pace, the standard of perfection had risen enormously under Gup's direction. Every gram underwent scrutiny, measure, and accounting in a tablet they called "The Book of Weights." The white beadboard, once used to build Emily's wings, had come to look like animal fat to their eyes. Gup hurled chunks and shouted, "Too hunky!"

Every piece had to be lightened. Thinly sliced aluminum parts fell under the drill. Engineers machined holes into nearly every major metallic device, creating a Swiss cheese effect that introduced a new aesthetic to airplane design. Under Gup's direction, the UROPs glued and sanded balsa strips on a thirty-two-foot table they'd spent a month building. Their new work surface provided an astonishingly level plane—perfect to within 1/32nd of an inch in its length and width and 1/10th of a degree in twist—on which to piece together the new wing sections for Daedalus. They needed better materials, Gup said. The ideal was weightlessness.

Gup looked more like a hip uncle than a colleague among his young crew. He'd left Lockheed in California late in the spring, a few months after Glenn's record flight. He had come back East with every intention of finishing his undergraduate degree and starting a master's program, but he seemed most devoted to building the world's finest HPA. Tall, tanned, lanky, blue-eyed, curious about everyone, still interested in even the most outrageous aeronautical ideas, he charmed the teenagers. He had a radical's disposition, an appealingly intense focus on life outside the academy. He was a rock climber and a sailplane pilot. He jogged and cycled and listened to New Wave music. When he opened his apartment for kids to crash, they found shelves of heavy academic books on aeronautical theory, airplane structures, and physics. But among these bricks they saw Madeline L'Engle's classic *A Wrinkle in Time*, and Tolkien's masterpieces of fantasy. Even as he entered middle age Gup seemed like a suitable mentor to the youngsters, offering great possibilities with razors, glues, balsa, and Styrofoam. He exceeded their imaginations and challenged their skills.

Out went the white beadboard. Gup fell silent, finally, turned, and stood breathlessly in the doorway. He smiled.

"Look," he told his workers, who continued to stare disconcertedly around the solar home. "If we get this airplane under sixty-eight pounds, I'll buy a pig and roast it."

Juan and Gup made odd roommates. Their apartment in Watertown occupied the entire first floor of a spacious old New England saltbox. But Gup's personality showed in almost every room, where he'd carefully hung pictures, maintained tidy bookshelves, and decorated with a few antiques. He kept the kitchen clean and well stocked. Guests noted Gup's sense of order, and commented on his good taste.

Juan, on the other hand, was accustomed to traveling light and living unencumbered. As usual he laid a mattress on the floor of an unadorned bedroom, tended a small bookcase, and lived like a bohemian. At least ten years younger than Guppy, Juan had devoted so much of his time to the single-minded pursuit of airplanes that sometimes it even annoyed Gup. He feared Juan had retreated into a specialty, a dangerous proclivity among MIT graduates, and would suffer for it some day. He sympathized.

"It could be a problem," Gup would say. "What Juan needs to do is work with his communication skills. To some extent this project helps a little bit because he grew a lot in working with other people. But that's something he needs to do even more, because he's a good engineer already. To some extent it's true of me, too. I don't like bureaucracies. I like a project where you see what you need to do and go ahead and do it. I sometimes worked like that at Lockheed, even when it wasn't the way to do things. And most of the time, it was okay. I got tolerated."

By midsummer, Juan was spending less time at the hangar. More and more, he relied on Claudia Ranniger, one of the UROPs he'd trained to build Emily, for wrapping and curing new carbon tubes. His disposition lightened. He became more sociable, and saw friends outside the project. But he remained absorbed by the airplane and increasingly interested in mythology. When he slipped into the Hanscom hangar occasionally at night to finish chores or to walk outside under the stars to forecast the weather, he felt a distant kinship.

One hot night, shortly after midnight, Juan wandered into the hangar to inspect Emily before making the weather report. With months of work ahead to finish the Daedalus airplanes, the team had been forced to retrieve Emily from the Air and Space Museum and truck it back to Hanscom for test flights. What Juan and Parky had come to realize, even before the record flights in California, was that Emily was built *too* strong. From data gathered on the lakebed, they saw that Emily wouldn't have made it even two thirds the way across the Daedalus route. Built originally to withstand the pressures of a 2,000-foot cliff launch from Crete, Emily had the strength to handle pressures three times the weight of gravity (three Gs), and as a result the airplane weighed far more than was necessary.

They'd redesigned Daedalus for launching at sea level, and the new frame would only allow 1.75 Gs. By giving up structural rigidity, Juan would take a calculated risk: the new Daedalus could be made to handle a few test hops with a team of pilots, and take one seventy-mile trip across the Aegean. But no more. The engineering aesthetic required Daedalus to be only as sturdy as a dragonfly, as evanescent as a cloud. This isn't the airline industry, Juan would say. Daedalus isn't a "product." Daedalus is a dream.

Late that night, Juan walked up and down the hangar beneath Emily's wings. He seemed nostalgic, yet entranced by the idea that the next plane they built would take them even closer to the edge of impossibility. He reached up and touched the Mylar skin as he paced.

"We realized that Emily was almost indestructible," he said. "So we can use Emily for training pilots and doing these test flights. But the new graphite for Daedalus is much lighter. It'll be much easier to destroy. You'll only be able to hold the tail boom near the front. It won't be engineered to be held by human hands. Daedalus is made for one purpose—to fly from Crete. Otherwise, what good is it? After the flight, it's sort of meant to die."

While Gup set the standard on the line, Juan's dictum guided them—"Simplicate and add lightness." Building an airplane that would fly nearly twice as far as Glenn's record distance in Emily required a new level of dedication. Recreating the Daedalus flight demanded paradoxical thinking—simplify while you complicate, add as you take away. It was ingenious. It was thematic. The decade-old lessons of MIT's radical rocketmen had passed on in this hangar from Gup to Juan to a couple of dozen MIT undergraduates looking for adventure and a break from the questionable utility of academic theories. Efficiency, aesthetic design, TLAR (That Looks About Right), elegance, the use of available materials—the canons of old rocket hackers would lead the construction team on a quest toward pure engineering. "We have no constraints now," Juan said. "If we're eating at McDonald's, you look around all of a sudden and there's somebody inspecting the Styrofoam containers, saying, 'How could we use this?' "

Juan pointed to Emily's spar. He'd dropped the "three kissing tubes" design for a more conservative design on Daedalus. The discovery of a higher grade of graphite, used primarily in satellite technology, would allow him to cut structural weight significantly.

Under the transparent wing, he examined Emily's bones and tendons. The twelve ounces of Kevlar threads, used for cross bracing in the wings, would be cut to four ounces on Daedalus. Even the plastic Mylar skin would be replaced on the Daedalus tail: one-half-mil thick on Emily; one-third-mil on Daedalus. They'd exchanged 1/64-inch plywood for twenty-millimeter basswood to produce lighter supports on the Styrofoam ribs. Out in San Jose, Parky used $10,000 from MIT to buy shop equipment and turned his garage into a lab to build lighter gearboxes; they would drop from 3.5 to 2.5 pounds. Juan wouldn't abide so much as a mistaken drop of epoxy. The builders even had to factor in a possible seventy-one grams of sponsors' decals on the rudder. Juan became the stern keeper of the weights, and accounted for every ounce in "The Book of Weights." On one page of the book, Mark had circled a coffee stain and noted: "0.2 milligrams."

Looking at Emily, Juan shook his head. "All bashed to hell," he said. The Mylar skin sagged slightly under the wing, the trailing edge had warped. He stroked the underside of the wing and found a couple of broken ribs. "Emily looks really old. It's a 1986 airplane now."

More than any member of the core engineering team, Juan seemed the least inclined to wax poetic about their endeavors. Not everyone saw this side of Juan, as he often masked his feelings with a tart, even cynical wit. But late at night, he spoke of the way the myth of Daedalus had made his own yearnings real.

"I've been into airplanes and things that fly since high school, okay?" he said. "Some people really do have that dream of flight and it pushes them on. Well, that's my dream, see. And there were these people thirty-five hundred years ago in a totally different culture and a totally different part of the world who had the same idea and thought it was great, too. But people thought they were crazy, and that this idea was impossible. Now we're going to show that this wasn't just a silly drama that somebody made up. We're going to prove that this can really happen."

Juan settled down with a book of ancient Greek history for a few hours that night, and at 3:00 a.m. he marched out to the runway carrying a stick strung with a Mylar streamer. Under the stars, he held the stick high in the air. A slight breeze stirred the streamer but didn't lift it. "We'll see," he said. Forty minutes later, he double-checked with the airport tower and decided to call the team out for flight ops. Morning would be windless.

For almost an hour, he operated a phone in the office headquarters, dialing every UROP and engineer on the list.

"Get up, asshole," he'd say, and snicker. "We're flying."

* * *

The old stories of Chrysalis, Monarch, and Emily were routinely repeated among the undergraduates. The Finberg legend grew as students from the electrical engineering department found themselves apprenticed to an eccentric electronics hacker who eschewed microprocessors in favor of building circuits by hand and labeling electric wires with Morse code. "*Finberg!*" became one of the most commonly heard remonstrations in the hangar. His electronics crew, cohorts of a generation that knew how to program computers but had never learned the pleasures of solder and circuit boards, honed new skills. Unfortunately, classes were held on Finberg Standard Time and the master's devices were sometimes so intricate that they didn't work.

"He arrived here as a child, and appears human because he was raised by humans," the South African team member, Grant Schaffner, would explain to fledglings. "The only time his ideas don't work is when he can't find the right parts on this planet."

But most of the tools they used were stored in a four-shelf supply cabinet—tongue depressors, paper towels, clothespins, X-Acto knives, C-clamps, sanding blocks, string, levels, an array of epoxies and glues, paintbrushes, spray adhesives, and razor blades. Their work depended less on the electric belt sander, jigsaw, and drill press than it did on the commonplace items that a child might use to build a balsa model—MIT engineering as they'd never imagined. Their work days lasted twelve, fourteen, twenty hours.

Gup took them swimming at Walden Pond and bought a grill so they could eat lunch together outdoors in the afternoons. As the head of the engineering team, he wanted to create a spirit of community and mission. He had clear ideas about what the project signified and he talked effortlessly about motivation.

"Here you create your own little empire," Gup would say. "That's what it's really about, I guess. One of the neat things about this project is it's outside of it all. It's not the usual rat race. I guess we're concerned about just getting it all done and not making any mistakes—engineers always make mistakes. But it's also about using as many clever ideas as you can and being as good and elegant as we can. Those are engineering principles—measures of goodness. To some extent you measure yourself by those standards, and here's an opportunity to measure yourself again and again and again for months and months and months. . . .

"Then, there's no commercial needs, either. It doesn't have anything to do with scheming to put bread on the table. And once the plane's flown, it's over. It doesn't fly anymore. . . . The important thing is the journey, not where you end up. That produces a sense of open-endedness or purity, I guess. It doesn't have any other purpose beyond that."

Visitors who entered the project's offices and shops often puzzled over the frenetic activity. Out in the solar home they'd see cans of corn, boxes of cereal, a toothbrush. A mattress pad laid on the deck behind the shed where Juan and Claudia built carbon tubes. Naptime occurred at odd hours.

The UROPs tuned in to Gup's dreams. ("I'd prefer it without a hundred and fourteen feet of wing on top of me," Gup would say, "but if that's the only way in this world to fly, and you have to be really tricky about it, well, that's it, baby. I'm in.") Tom Clancy dropped out of school and devoted all his hours to Daedalus. Grant stayed at Gup's right hand, working shirtless, often with only a strip of sandpaper, shaping wing ribs by hand, making pieces so they'd please the eye. And Dari Shalon, an undergraduate who'd helped build Monarch in 1984 but who had left MIT to travel abroad, returned to Hanscom with a sense of devotion that sometimes seemed to be an obsession. He brought the music of Vivaldi and Mozart into the hangar.

In his travels through Europe, Dari had stayed in touch with John, and at one point, he'd even bicycled through Greece and toured the islands of the Aegean looking at possible takeoff sites for Daedalus. He'd studied Daedalus mythology in Athens, and sent letters to John with suggestions for a flight path. That had been more than three years ago.

When Dari returned, he stopped first at Hanscom just to visit old friends. But seeing the airplane, he was swept away.

"When I saw them building Daedalus," he said, "I realized they weren't creating a functional airplane. They were making a work of art." He returned to MIT as an undergraduate in the mechanical engineering department and immersed himself in the Daedalus engineering team. More than anyone, Dari became absorbed by the aesthetics of the airplane.

One day in late July, Dari walked into the hangar and saw the new material that would replace the white beadboard. It was much smoother to the touch. Mark had insisted that the wings needed a silkier texture because the old board's slightly beaded surface created too much drag. The construction industry, they'd discovered, had just produced a new extruded polystyrene insulation board with remarkable strength and a tight, closed cell structure that made the surface utterly smooth. Known as Foamular, a thin sheet would make the world's cleanest covering for the wings. Unfortunately, it was pink. Dari looked at the blocks of Foamular and shuddered. Pink wings?

"This doesn't fit with what we're doing," he said. "Pink doesn't go with a black spar. This will look like a little baby girl airplane."

He spent days combing supermarket shelves around Lexington and Concord for the right shade of food coloring to match their golden Kevlar twine. The best match was a lemon yellow, but sprayed from an atomizer on the pink polystyrene, the colors combined and turned gold.

One night, Dari worked alone on a sample block of Foamular. He sprayed it gold, then set it on a table next to a pink block the same size. He ran out to the hangar and called the other UROPs over to look at what he'd done. "Keep an open mind," he said, as he led them through the dark. When he opened the door, both blocks on the table were pink. Dari suspected a prankster at work. He sent them back to the hangar and tried again.

The second time he dyed a piece, Dari stood by the table and waited a few minutes to see if anyone was watching. As he waited, the wet dye evaporated and the gold piece turned pink again.

Dari spent days searching the Boston phone books for dye manufacturers. "Look, this is an emergency," he'd say, calling potential clients. When he finally found a chemist at one company who seemed to take him seriously, he mailed the pieces Federal Express.

A few days later, a call came in to the Daedalus Project office. A chemist asking for Mr. Tidhar Shalon.

"Are you the one with the airplane?" the chemist asked.

"Yes," Dari said.

"You wanted a gold airplane?" he asked.

"Yes," Dari said.

"Take it from twenty years of experience," the chemist said. "Forget it."

Not long afterward, Dari tried to persuade his colleagues that they should apply a non-mathematical concept to the Daedalus construction: the Ugliness Drag Coefficient. Whatever pleases the eye, he said, pleases the air. It took weeks for them to convince him that pink wings would fly.

By August, two major sponsors had signed contracts with Daedalus. The president of the Shaklee Corporation, a nationwide health and nutrition company, tentatively agreed to spend $150,000 on the selection, training, and support of the pilot team. A certain amount of money would go to Ethan Nadel, as well, to develop a sports drink, which seemed increasingly important for a four-hour flight.

Then, United Technologies Corporation generously provided $400,000 for the construction of the Daedalus plane and its test flights in California. United Technologies, maker of the Pratt & Whitney engine and Sikorsky helicopters, became the primary sponsor after months of careful negotiations led by Jack Kerrebrock.

Donations of materials followed from Amoco Performance Products, Hercules Aerospace Company, IBM, Mitsubishi, Nikon, Polaroid, Sikorsky, and a score of other corporations, from a plastic wheel manufacturer in Walworth, Wisconsin, to a company that built electronic heart-rate monitors in Kempele, Finland. The Charles Stark Draper Lab in Cambridge gave Steve Finberg and Bryan Sullivan leaves of absence to develop avionics and a flight control system so Daedalus could fly at night. Lois McCallin, who had continued to miss her friends, sold her boss at Fidelity Investments on the idea of letting her join the project's public relations staff, at least temporarily.

As the summer passed, the Daedalus Project continued to expand. John held more staff meetings, traveled more often, typed out a flurry of memos, letters, contracts, and schedules. There was the first of many flow charts. The office staff took over a couple of hundred square feet on the second floor of the Hanscom hangar. John stocked it with phones and Macintosh computers, hired a woman in Athens to open a Greek branch of the project, and developed a computer mail system to communicate with her and Parky in California. John came to work every morning an engineer, and within half an hour he was deluged with the nuisances of an office manager: a reporter from *The Wall Street Journal* wants to visit; a decision has to be made about the flight route; United Technologies doesn't like the team's logo; Barbara Langford is feeling abandoned in Washington. . . .

John developed a nervous tick—a blink that crinkled his lids as if he'd been hit between the eyes with a slight charge of electricity. But no one could keep up with him when he walked, and as he talked on the phone with sponsors or supporters or friends, the most common words were exclamations, "Great! . . . Neat! . . . Great! . . . Great! . . . Great!" He seemed to handle the pressure with a kind of innocent, southern charm and a proficiency that made his employees feel secure, at least for a while.

To reporters, he offered an image of the project that promoted his teammates and kept them from focusing simply on the pilot or a single builder. "We live in an age where the myth of the individual is supreme in our media," he told *The New York Times*. "I can't point to a single researcher or pilot who has made this project possible. It's a collective dream for design engineers, physicists, physiologists, classicists, athletes. . . . In that sense, the project represents the spirit of Daedalus, who was an artist and a scientist at the same time, a person who combined in himself the disciplines of many."

John would struggle again and again with this self-created dogma. He made a rule that no single person could have his or her picture taken alone with the airplane (although he violated it once, for *People* magazine). He controlled the project's image to keep it pure and simple. Even as he talked with documentary producers about making a film, he tried to explain that while the scientific and cultural aspects of the project were certainly important, the significant underpinnings came from childhood.

"You have completely missed the real story of what is going on in the Daedalus Project," he told a producer who'd proposed a piece for the public television series NOVA. "I believe the story goes more like this . . . Once upon a time, there were three friends—Bob, Guppy, and John. All had built models since childhood, all entered MIT because they liked things that fly better than anything else. Rockets, gliders, airplanes, it didn't matter. . . ."

There'd be no solitary hero in his Daedalus revision, no hierarchy. The engineering team of Langford–Drela–Parks–Youngren–Finberg–Cruz had been one of his dearest dreams, a creative core he wanted to transfer into a business after the flight. A gang of romantics and renegades—Daedalus, he thought, would at last give them a way to free themselves.

Reporters would listen to his spiel, puzzled by his intensity and earnestness. They'd ask what he expected to do when he returned to, as they would say, "the real world."

The question irritated him. But he would smile. Only people inside the team could really understand. "This is the real world," he'd say.

The external affairs branch of the project grew steadily under Peggie Scott's authority as she answered increasing requests for interviews, pictures, articles, explanations. In time, the team's heightened profile excited the public relations office at MIT, and plans were made to use the project to woo money in the next alumni fundraising campaign.

Institute Professor Phillip Morris made a speech at the Council for Arts on campus and called Emily the most beautiful airplane in the world. "It is wonderful," he told an audience of administrators, professors, and students. "It will not make a penny or drop a bomb on another hapless target—it is an insightful combination of hand and mind that makes a thing like that." *National Geographic* produced an eighteen-minute show about the flight of Emily. *Rolling Stone* mentioned the flight, too, remarking that Glenn's thirty-seven-mile record was "a lot further than anyone got on a scheduled airline this year." *Popular Mechanics* published a cover story about the "88-Pound Pedal Plane." Glenn Tremml received an invitation to have dinner at the Reagan White House, which he declined, because John decided that an individual appearance did not suit the image of their quest.

The Daedalus team seemed confident and assured on those sunny days in late summer. But one afternoon in late July, as Glenn sunned himself by the runway and UROPs toiled in the solar homes, a Yale classics professor came to visit. She carried a thick sheaf of papers, weather charts, complex records and maps. Dr. Sarah Morris knew the Daedalus myths backward and forward. At John's invitation, she would help them plan a specific route. Dr. Morris didn't bring a lot of weighty academic advice, but she did have news that would upset the balance between art and engineering. Her belief that the new Daedalians could have even more freedom posed a challenge not every one among the team was willing—or able—to accept.

16

All activity ceased when Sarah entered the hangar. UROPs left their posts; the office crew abandoned the second floor; Juan posted a sign on the solar home—"Go away!"—and lit out across the pavement. Two dozen crew members crowded into the conference room, jabbering expectantly as a magnetic young woman balanced two and three conversations at once and managed to hold off most queries until John stilled them with a shout.

Sarah shared their enthusiasm. As an associate professor of Classics at Yale, her academic research on Daedalus spanned millennia; she had collected dozens of versions of the myth, a traveling gallery of paintings, illustrations, sculptures, and engravings. While she looked pale and scholarly, she spoke with an intensity that drew crew members close to the table, as much to watch as to listen. Sarah, a friend of Ethan's, offered the credibility they needed to make the correct choice. She was a flurry of esoterica, a burst of intellectual energy.

But where would they fly? they asked. At that, even Sarah ducked. Myth making is not so simple.

"I might as well introduce the question of the route," she said, "by reminding you of the way ancient mythology worked—at least in the classical world. There was a new version of Daedalus every decade, every time a new art form was invented. It's very hard to ask for an orthodox version of the myth . . ."

"No!" Dari interrupted. He was as insistent as he'd been about the pink wings. "If there was just a source that we could track down."

John unrolled a set of maps on the long table in front of her. Sarah sighed, and glanced away. She would have to start again.

Her lecture became a sudden blur of Daedalus tales, from the fifth century, the fourth century, the eighth century. Every region and era produced its own version, she said. In some cities, Daedalus was known only as a great sculptor; at another time, he was the world's greatest metal craftsman; in a different culture, he was strictly an ingenious hydraulic engineer. In fact, she said, Daedalus had not always been famous for building wings. The Greek myths were filled with stories of winged creatures, flying horses, gods, and griffins. The idea that Daedalus built wings to escape the labyrinth was probably of such minor regard during antiquity that in some versions the "flight" was taken to mean an escape by boat.

Not until Judeo-Christian times, after the myth of Daedalus merged with the story of Icarus, did the idea of human flight become the significant theme. The invention of wings—the first evidence of human flight—emerged through the mythologizing of moralists who used the story as instruction for children to obey their elders and as a cautionary tale for those who would tempt fate. Human flight disturbed Christian theologians, who used the death of Icarus to deliver warnings against pridefulness.

Aside from modern times when the Christian Church promoted the Icarus-Daedalus version, one theme had been constant. "He was always flying to get away from something," Sarah observed. "He was always trying to get away from King Minos."

By shifting the topic from the route, Sarah distracted the team from its compulsive mapping. She passed out a collection of pictures showing Daedalus artwork, as it appeared on vases and in paintings for more than thirty-five hundred years. Juan, who'd become the team's authority on aspects of Mycenean culture, popped questions and wisecracks at her about the lowly status of engineers in ancient Greece. Dari piped up again about the route. Sarah listened and then turned to John, who was also gazing at the maps.

"We want a simple, chronological story that goes like this: 'Father . . . mother . . . son . . . the career . . . ,' " she said. "But you can't do that. I'm sure if there was to be a modern version, Daedalus would be designing a human-powered aircraft. You should be thinking of your work as the last version."

John patted the papers stretched out before them and straightened their edges. Dari folded his arms and looked away. Sarah scanned the room, still full of expectant faces, then laughed. She had tried her best to challenge their imaginations, but they'd had enough classicist hopscotch. John took over.

"To refresh everyone's memory," he said, "there are two major routes under discussion."

The first route he plotted originated in the seaport city of Iraklion, on the island of Crete, more than one hundred miles from the mainland. (See map on page vi.) Daedalus had to fly from there because Crete was the home of Minoan civilization. The remnants of King Minos' palace stood in Knossos, just a few miles outside of Iraklion, surrounded by vineyards and olive orchards. The palace had been discovered at the turn of the century by British archaeologists, who uncovered in those rocky fields a sophisticated network of ancient pipes and an intricate architecture. A series of hallways and stairs led to bedrooms and servants quarters and passages that angled toward an altar, a theater, tiny storage rooms for holding wine and food, and a central arena, apparently used for celebrations. Built like a maze, engineered by a deft mind, the site at Knossos was reputedly the home of Daedalus in 1600 B.C. Along the walls of the palace, archaeologists also had discovered paintings of double axes, called labrys, the root for the word "labyrinth."

John, Peggie Scott, Steve Bussolari, Ethan Nadel, Sarah Morris, and a few members of the team's meteorological crew had each visited Crete. John and Peggie preferred Iraklion as a takeoff site because of the availability of good hotels for the press and the proximity to a Greek Air Force Base, where they could make practice flights, enjoy security from curiosity seekers, and harbor from foul weather. Under their plan, Daedalus would fly from the Air Force Base to the island of Santorini, about seventy miles away.

The other route, favored by Bussolari, began near a remote village in western Crete called Maleme, and actually ended on the mainland of Greece, in the town of Neapolis, at the southernmost tip of the Peloponnesus. There would be no hotels at either point, no lodgings for the press, no attractive restaurants. But there was an abandoned airfield near Maleme and a soccer field that would make a dandy takeoff site.

John turned to Sarah, hoping for support. He thought the choice was obvious, but the classicist would not be pinned down.

"Keep in mind, John, that there were several routes," she said. "So it depends on which version you want. Ovid's version mentions that he flew over Santorini, of course. But there are also versions that mention the island of Ikaria, a flight from Athens. This second route, from western Crete to the southern Peloponnese, is a very interesting one in the history of technology because Minoans traveled in the middle of the third millennium and settled on. . . ."

Bussolari spoke up. "What about Santorini?" he said.

As it turned out, Sarah was not too keen on Santorini, either. The island was spectacularly beautiful, shaped like a horseshoe, shaded with dark, muscular volcanic cliffs. It was a caldera, the startling leftovers of an eruption that collapsed the volcano about 1500 B.C. and destroyed settlements miles around with fire and tidal waves. But in recent years, an archaeological site on the southern coast had stimulated a lot of publicity—"damaging publicity," she said—linking Santorini with the lost continent of Atlantis.

"I happen to think it has plenty of publicity potential," she said. "In July and August, there are probably forty thousand people on this island. It's a very small island, and I myself don't think it needs any more publicity because it's been overloaded with this sort of Atlantis . . ."

John, scowling, interrupted again.

"We saw slides at the last staff meeting of this route from western Crete," he said. "There are no roads there, or they are lousy, so the best access to the takeoff site is by boat or helicopter. And when you start talking about maybe staying there for sixty days with an airplane and a staff of maybe thirty people, and you can't get anything there but by helicopter, it begins to sound insane to me. It begins to look very much like going from nowhere—really nowhere—to almost in the middle of nowhere."

Bussolari leaned over the maps and stretched uncomfortably.

"Well, let me say something," he said. "The logistics between Maleme and Iraklion are much closer than what John's said. The only difference is lack of hangar space. Maleme is much quieter and we could have Maleme to ourselves. Iraklion is a very busy municipal airport. . . ."

Members of the team began to whisper.

"About how much longer is that flight?" Juan asked, pointing at markings on the map that led from Maleme to the mainland.

"About five," Bussolari said, scanning his notes.

"Five miles longer?" Juan asked.

"It's about an eighty-one-mile flight from Maleme to the mainland," John said.

"And the other one is seventy-two miles?" Juan asked.

"And the other one is seventy-two."

"So the distance is negligible," Juan said.

There was some uncertainty about distances—the translation of kilometers to miles blurred in the discussion. It was a critical point for the physiologists.

Bussolari and Ethan had discussed the two routes already. The extra distance from Maleme to the mainland might tack nearly thirty minutes onto the flight. "It's hard to predict," Bussolari said, "but we felt that extra half hour might represent a significant difference in terms of the pilot's ability to make the flight. Any difference in flight time less than a half hour, though, still represents the same flight. Once you get above a half hour, you could say there's a slight chance that . . ."

"That the pilot could have made one and not the other." Glenn Tremml spoke up so suddenly and yet so quietly that Bussolari did not even seem to recognize the voice.

"That's right," he said. "It's not the point where it would be possible versus impossible, but that's the point where we would start to curl our brows over this."

Glenn tried to speak again, but Dari moved away from the wall toward the table, waving his hands.

"We have to make a basic decision about which route we want to take and the logistics should fall behind that decision," he said. "But if you transcend the logistics for a second and think about what we're working to recreate—and that is the myth—then I think this western route has a definite advantage because you're flying from Crete to the mainland. You're not flying from one island to another. And this route from Iraklion to Santorini is from a port to a resort."

The phrase hung in the air. It was repeated around the room. There was the sound of startled laughter.

"But the western route," John said, finally, "it's nowhere in Crete to nowhere on the mainland."

"That's the whole point!" Dari said.

Chattering overtook the room. What had been a cautiously circumspect discussion broke into debate. UROPs and lead engineers weighed the value of flying from nowhere to nowhere against the specter of flying from port to resort. Mark argued that he didn't see why they should invite the press, anyway. Dari said theirs was a mythical quest, not an ad-venture. Glenn still struggled to be heard above the din.

"In my opinion," Glenn said, "in my opinion, if you push the pilot too far, he may deteriorate for ten minutes, but he won't believe it until fifteen minutes later."

What did this have to do with the myth?

"So," he continued, "it's really only five or ten minutes in which the pilot might say, 'I'm in big trouble.' And if it's humanly possible, he might be able to hang in another ten or fifteen, if his life depended on it."

"Right," Bussolari said. "It's just enough time to get the airplane under tow."

"That's the way it works in the lab," Glenn said. "If Ethan says, 'I need to get another measurement, can you hang in there?' the best you can do is five or ten minutes."

But Dari intruded again. The issue was too important to be sidetracked by the pilot's insecurities.

"Look, John," he said. "I see this one, from Iraklion to Santorini, as a sellout. You're selling out to (a) the press, to make it more convenient; and (b) yourself, to make it more convenient, to be able to live in a hotel or not."

"I didn't say that," John said.

"No, wait a second," Dari said. "If we did the route we wanted to do, and if we're working so hard to recreate the myth, I'd rather see a picture of the plane flying out along a cliff on the western coast than landing between a retired couple from Miami Beach on Santorini. That's the kind of picture I'd like to see."

Peggie charged in. "Santorini's not that crowded," she said. She understood the team's sentiments, but she also had to satisfy their corporate sponsors, find lodging for the press and the team, and work with John to overcome countless logistical nightmares.

"The thing is, I don't see a problem with setting up a base camp up here in western Crete," Dari said.

"But, Dari," Peggie said, "what Dr. Morris is saying is we're not necessarily being true to the myth by going . . ."

"No, no, no!" Dari said. "She didn't say that. She said either one is just as true."

"But, but, but . . ."

Shouting broke out. The engineers felt confident anything was possible. They clung to a vision Bussolari offered, accessible by goat paths, out of the eye of the media and the glare of commercialism. Dari motioned at two points on the map: The western route, nowhere to nowhere.

When John called for a vote and hands were counted, it was apparent that the new Daedalus myth had become a flight from nowhere. Besides Sarah Morris, who chose not to vote, the only other abstention was the pilot. Glenn Tremml laid low.

The moon hung over Hanscom even as the sun rose. The sky bled handsomely, and hangar doors slid wide. At 6:00 a.m. four students escorted Emily down the runway for the third beautiful August morning. As they trotted, the airplane lifted off the ground, taking flight in their hands. They fussed over the wings, waved spectators away. Gup and Juan sprayed the prop with a chemical mixture of kerosene and dark powder to test the new blades. Emily looked angelic, glistening pristinely at sunrise.

No one noticed the troubled engine. Glenn, who had stripped down to a T-shirt and running shorts, was not smiling beatifically for the *National Geographic* photographer.

He had already flown half a dozen times up and down the runway that week, testing an improved water system, deploying a new towline. He'd worked in the office every night until dark, and slipped out of bed at 3:00 a.m. each day for flight ops. Over three days he'd slept a total of seven hours, and now the engineers expected him to make at least another eight flights. Glenn wondered whether the team remembered that Daedalus was a man, not merely a technologically advanced airplane. Increasingly, they referred to Glenn as "the engine." But he was beginning to have doubts. Not long after the meeting with Sarah Morris, Glenn had entered a triathalon in Middletown, Connecticut, and found himself lagging behind the leaders. Even worse, ergometer tests at Yale had identified a few more pilot candidates—top-ranked national cyclists whose test results topped Glenn's best. In Athens, Ethan had exacted pledges from a couple of cyclists on the Greek Olympic Team whose average power outputs exceeded Glenn's by more than 10 percent. By the end of the year, he realized, at least two more Americans and a Greek would compete with him for a place in the cockpit. His best odds of becoming Daedalus had narrowed to one in four.

The flight to nowhere, which had caught the engineering team's fancy, distressed Glenn. Whenever he tried to talk to Dari or lobby the UROPs, they offered condolences. Not even an Olympic athlete, he believed, could do what the team now expected.

Just the night before, after the engineers and office crew went home to get some sleep, Glenn unburdened himself to a news reporter. With Daedalus in its last stage of construction and flight ops scheduled in California during December, all the engineers needed were a few trials with the prototype to test the new propeller, water system, and tow device. They'd gone home optimistic and happy. But Glenn felt betrayed.

"You see," he explained, "what happened was they asked Ethan, 'What would be a significant difference in distance? I mean, if you were dying, you know you could make it another fifty feet, right?' But he says, 'Five miles.' It was just a number he pulled out of the air. . . . And people just extrapolated from that. He pulled a number out of the blue and it didn't seem significant to anyone. . . . I thought about it, and then I said, 'Well, what if we take the longer route and I put it in the water during the last five miles?' And Dari said, 'It's all in your mind. You won't do that.'

"But see, I've got a good friend who tried out to be the pilot for this project. He's quite a fierce competitor, too. And he quit on the ergometer at three hours and fifty-eight minutes. Here's a guy who knew if he could stay with it for just two more minutes, he's on the team. And he gave out. How do you explain that? It's not psychological. It's physiological.

"So what are the odds of just falling apart if I get that close? I'm not saying it can't be done. I'm not saying I don't have the confidence. But they just keep telling me, you can go for six hours. And all I think is, I don't know."

The engineers' attachment to the flight from nowhere certainly seemed beyond comprehension. The original estimates of a sixty-eight-mile flight had meant Glenn would pedal four and a half hours over the Aegean. But if they chose the eighty-one-mile flight out of western Crete, the trip would take at least another hour. And questions about possible head winds had never even been addressed. Glenn, who knew more about the limits of endurance exercise than any of the engineers, had serious doubts about chancing fate for up to six hours, and he couldn't forget that he would also be the only pilot who never passed a four-hour test.

But despite his doubts, when he climbed into the cockpit wearing a bright blue and gold T-shirt that read: DEPENDABLE ENGINES (a logo of United Technologies' Pratt & Whitney engines division), the words puffed out from his chest. He looked self-assured on takeoff, and flew silently up and down the runway. Between flights he talked enthusiastically with visiting reporters and photographers, spitting out more pablum than a press release. In two minutes he could explain to visitors how, in the best tradition of interdisciplinary research, Daedalus had introduced classicists to technologists, acquainted physiologists with basic engineering principles. He gave a breezy presentation that allowed the team to think he felt just fine.

The engineers and UROPs were strewn up and down Hanscom field like a traveling circus. Gup tooled around on a recumbent bicycle. Lois served chocolate doughnuts for breakfast, just like the old days. Between flights someone hung a miniature pine tree air freshener in the cockpit as a joke. A UROP pinned the Styrofoam Finberg head—IDOLBERG—atop the altimeter for good luck. As Glenn stooped over to take his seat for the final flight, Juan leaned in. "You're brave," he said. "You fly without plastic Jesus." Juan winked and laughed.

After eight flights, the engineers had photographed, videotaped, and analyzed the chemical streaks that indicated a transition line of turbulence on the prop. The design, they agreed, was flawless.

The in-flight tests officially ended for Emily. By December they'd have the first Daedalus airplane ready for flights on the California lakebed. At 8:30 a.m., they hustled Emily off the runway for the last time and rolled her back into the hangar. With the airplane secured, they bounded for their cars to celebrate with a big breakfast at Denny's Restaurant. Glenn dawdled on the runway with a reporter and made them wait. He was talking about the day Sarah Morris had set them free. He was thinking about the flight to nowhere.

"Basically," he said, "the engineers will be finished by October. The plane will be built. There could be some minor modifications and improvements. But in the end, it's the pilot who climbs in and either makes it or doesn't. Sometimes it seems like they realize it. Sometimes they don't."

He was still slightly breathless from the last flight. He was sweating, and as his chest rose and fell, the DEPENDABLE ENGINE logo expanded and deflated. As he walked slowly back to the hangar, he pulled off the blue and gold T-shirt. Underneath, as dark as doubt, he was wearing a new shirt, almost lustrous in its blackness. In big letters, covering half his chest, the logo said: NO PROBLEM. And in tiny script near his shoulder, as simple as a wish, as thin as a thread of Kevlar fiber, it said: "*Santorini*."

Ethan enjoyed this kind of challenge. The dynamics on the project worked less like a support group than a game of Russian roulette. When the Yale professor's time came to spin the cartridge, his pulse quickened with pleasure. Reporters would call, after talking to Glenn, and question the pilot's analysis. If a four-and-a-half-hour flight would put an athlete on the edge of human endurance, how could they contemplate an extra hour? Ethan laughed.

"I'm sympathetic with Glenn's position, of course," he'd say. "And I'd like to bring some reality into the program. On the other hand, if we planned a trip that were thirty miles, it wouldn't be so interesting, would it?"

With less than three months left to select a pilot team, a call went out to the best amateur cyclists in the country. An advertisement, framed under a silhouette of Emily over Mojave, appeared in national special-interest magazines, such as *Triathalon Today* and *Velo News*, a cycling publication known among serious amateur racers:

> The Daedalus Human-Powered Flight Team is looking for national class endurance cyclists to pilot the Daedalus human-powered airplane from the island of Crete to the Greek mainland in celebration of the Daedalus myth. This five-hour, pedal-powered journey in the spring

of 1988 will establish a new world distance record for human-powered flight.

Requirements for members of the Daedalus flight team:

- VO_2 Max on cycle ergometer of approximately 70 ml/min/kg
- Capable of producing at least 3.3 watts of power per kg of body weight for five to six hours. (This is equivalent to .28 to .32 horsepower for a 140 to 160 pound person.)
- Pilot experience highly desirable.
- Time commitment required: August 1987—approximately three days of physiological testing at Yale University. November 1987 through May 1988—full-time commitment to physical training, pilot training, flight operations in Greece.

Nearly three hundred applications poured into the project office. John had to hire a UROP full time to assist Ethan with test preparations and correspondence. Several professional cyclists showed up at Yale to take VO_2 Max tests. One applicant wrote that he'd flown Japanese Zeros in Hollywood B movies. A few top-ranked amateur skiers applied because they had high VO_2 Max scores, as did a member of the Harvard crew team. Ethan invited more than two dozen athletes to laboratory auditions, first, to take the VO_2 Max test, and then, if they could produce power greater than Glenn's best score (3.3 watts per kilogram), they were asked to take the four-hour test. The project paid to fly them in and fly them out. They came from as far as California, as nearby as Boston. If they passed the first test, the athletes got a free meal at an Italian restaurant near the lab and a night's stay at the local Holiday Inn.

Ethan sometimes tried to predict by sight how a candidate would score. The failed ones often surprised him. One of the professional cyclists couldn't finish the four-hour test. Amateurs often scored better than the pros. Cyclists with the most impressive résumés did not produce the most power or prove to be the most efficient airplane engines. To Ethan's surprise, he began to learn a little more about the mysteries of physiology. The blood samples, the oxygen tests, the heart-rate charts became increasingly vital to his task.

On hot summer mornings and into the early fall, Ethan and Bussolari went for long runs through New Haven, returned to the lab soaked with sweat, and shared bowls of yogurt and granola sweetened with raisins for lunch. Then Ethan would stash his wet clothes in a corner and they'd enjoin a lab assistant to help them conduct tests on the ergometer. Erik Schmidt, a national-class cyclist, flew in from Denver to try out. Bryan Miller, a member of the U.S. national cycling team, left the Pan American trials to take a four-hour test. Sally Zimmer, a runner for the New England Adidas team, tried out. Frank Scioscia, a long-time national-class cyclist, interrupted his U.S. tour and came into the New Haven office exhausted from his travels, but determined to make the Daedalus team. During the four-hour test when Frank's legs cramped up, Ethan and Bussolari watched, astonished, as the cyclist hammered his thighs with his fists to keep the legs churning. The new pilot candidates were gruff, full of desire.

Eleven men made the cut for a four-hour cycling test, using a commercially available exercise drink to forestall fatigue. Three of the eleven dropped out of the duration test before the four-hour mark. Ethan had adequately estimated that a drinking schedule of one liter an hour would prevent dehydration by keeping the concentration of blood sodium and the plasma volumes constant throughout. But the athletes' blood-glucose levels plummeted by the fourth hour and their heart rates rose to levels that, if they had been forced to continue, would have led to failure. If the Daedalus flight extended much beyond four hours, Ethan realized he would have to develop a new supplementary drink to replace the lost fluids, carbohydrates, sodium, and potassium. Glenn's fears, it seemed, had been based in fact.

Back at Hanscom, though, when members of the engineering team heard about Ethan's experiments, they joked about Dr. Nadel's proposed elixir. Someone drew a cartoon of a medicineman's wagon with Ethan's name inscribed on the side and tacked it to a bulletin board. Scoffing at the rumors that the Shaklee sponsor might market the drink publicly after a successful flight, they labeled the would-be Daedalus drink "Ethanol," and wondered whether a maverick cyclist would eventually turn their flight into just another commercial sporting event.

17

Even under pressure, Glenn looked as tan as a buck eye, as calm as a cow. He took Grant Schaffner's girlfriend dancing at night, and drew pictures on the Macintosh in the afternoons. He volunteered to talk with news reporters and conduct research on heat conditions in the cockpit. For a while, he even encouraged Gup to think of the fuselage as a kind of flying fish, and tried to persuade him to cut "gills" in the cockpit as air ducts.

For two months after Sarah Morris's appearance, Glenn sublimated his fears. The engineers forgot about his predicament. In truth, the team's first American hero no longer felt comfortable questioning the route. With his scientific arguments rejected, Glenn decided to keep quiet and smile.

But in late September, after two other Americans and a Greek Olympian accepted jobs on the pilot team, a letter appeared on John's desk that revealed the dire nature of their pilot's predicament.

> To: Appropriate Team Members
> From: P. Glenn Tremml
> Re: Resignation
>
> I have made a very difficult decision, but one which I believe will benefit the project as a whole. I have reviewed my role and performance in this project and at this point feel that I am no longer the best person for the job. Therefore, at this time I have decided to resign as pilot/athlete for Project Daedalus. My decision is based on two factors.
>
> First, I do not believe that I have the given ability of many available athletes. Also, the difference between myself and these other athletes is significant and substantial. I have taken this physical challenge very seriously from the beginning and at this point believe the task

at hand to be impossible for me. This is based upon careful review of the literature, my best performance thus far, my physical condition since arriving in Massachusetts, and the recent increase in the task difficulty. For whatever reason, I have not been able to train as well as I did last summer while also working a full-time job, and am not nearly as fit. At this point, I should be at my personal best and still improving instead of trying to get back in shape. Indeed, I have serious concerns that any of the people tested can complete a six-hour flight, let alone myself.

Secondly, I feel that the project is not putting enough energy and emphasis on the human factors. The pilots on this project are professional athletes and should be treated as such. Things should be taken care of for them and very little should be expected from them outside their immediate job, which is to be the best prepared pilots and athletes for this nearly impossible task. Also, the research in areas of food, rehydration, thermal limits, human-machine interface are falling far behind. As we learned in California (last year), it's the little things that will kill us. I have found that many of the engineers do have an interest and concern for pilot comfort and have been very helpful, but each of them is already overworked with several jobs. As for myself, I have found that even being concerned with just two jobs, I have done neither one well. The situation seems like it will only worsen as the crushing deadlines, such as the rollout, appear closer. I feel that I can no longer play both roles as pilot and pilot advocate. I therefore wish to resign as a pilot.

Encounters between John and Glenn tended to be slightly formal, almost too professional. John read the letter and called his pilot in for a meeting. As usual, they were cordial, friendly to a fault.

"Basically," Glenn told John, "there's only one full-time person in the human factors area, and that's me."

"I don't understand," John said. "What about Ethan and Bussolari?"

"They're there, but what I'm saying is the human factors division spends all its time selecting pilots. So whenever there's been a pilot candidate, Bussolari goes to Yale to help in the lab and takes one of the UROPs with him. Well, I worked in that lab for three years and I know that not only can Ethan and his technician, Sandy, do it by themselves, but Sandy could do it even if Ethan wasn't there. So now we have Ethan, Sandy, Bussolari and a UROP."

John listened. "Okay," he said. "Go on."

"I just don't see how it was worth the time and effort to have them there. If there were other motives—for Bussolari to meet the pilots and represent the project—then it's legitimate. But I think it was just fun for them to go down there and be in the lab with Ethan and take blood and watch these guys."

"I don't think so," John said. "But even if that were true, that's got nothing to do with your training." He looked at the letter again. "I know Bussolari has lots of jobs. He has the pilot selection, the weather, and basically whatever's left of the human factors. So he's doing pilot selection and weather. And you're saying there's more? That still leaves the drinking system, the cooling problem, the pedal system . . ."

"I'm not saying it's all his fault," Glenn said. "But I've been up here for a whole half a year and nothing's getting done, except testing those candidates."

"Okay," John said. "I won't try to talk you out of it. What if I give you a new job?"

Glenn looked startled. He knew his only advantage on the new pilots' team would be his experience as a flyer. None of the other cyclists had ever handled an airplane before; few people in the world had ever flown a human-powered airplane before. With three new, superior athletes vying for a job in the cockpit, Glenn suspected there'd eventually be a scoring system to cut the lesser engines from the team. Ethan and Bussolari had already used his physiological scores as cut-off points for selecting the new pilots. He knew he already ranked at the bottom.

"I can't be a pilot and work on other things, too," Glenn said. "And now you're choosing guys who are genetically superior to me. No matter how hard I work, I can never be a match for this Greek guy."

"So," John asked again, "what if I give you part of Bussolari's job?"

No matter how he thought about it, Glenn could not envision losing his chance to fly Daedalus. He hadn't quit medical school to train other athletes. There was a pause, while he considered his predicament and gauged the risks.

"Of course we could always use you as a test pilot," John said. "Or we could reassign Ethan's UROP to do the thermal tests. And it looks like Lois will be available soon to handle some of these pilot training issues. . . ."

The negotiations lasted well over an hour. When Glenn walked out of John's office, he said he was once again a Daedalus pilot.

"I basically quit to jolt the system," he said. "And I did. I think John was pretty surprised."

But once he left the hangar, John only seemed relieved about having saved his most skilled pilot. He was glad to have kept a knowledgeable, photogenic medical student to represent the project, but he hadn't really understood Glenn's concerns.

The September negotiation resolved none of the significant issues about the physiological demands of the Daedalus flight. It only established a precedent for political jockeying on the pilots' team, and underscored John's belief that he faced a special puzzle with his engines. The performance of a human being, unlike a carbon tube or a slice of Styrofoam, couldn't be predicted. The so-called human-factors division would depend as much on psychological scrutiny as scientific analysis. It could become a manager's toughest job.

The management team John and Peggie Scott created formed an invincible partnership. Peggie matched John's enthusiasm and became his most trusted confidante. But while they struggled to maintain the ideals of the project—egalitarianism, decision by consensus, no commercial constraints—the truth was that they directed it day to day, touting its virtues publicly but privately giving the necessary orders to keep the project on track. Like the human-factors division, the management team often faced problems that had nothing to do with engineering.

By the end of the summer an increasing number of demands appeared around the hangar in the form of memos. The UROPs generally disregarded orders. Juan and Mark, who distanced themselves from the project over the summer, ignored most demands from above. But Gup, who oversaw construction, couldn't neglect notices to hit deadlines, no matter who signed off. During August and September, as the rollout of Daedalus drew near, he found himself in sudden confrontations with Peggie and the young man who had once held him in highest regard.

Beginning in midsummer, John's urge to direct the construction process irritated Gup. The decision to build two airplanes at once—they needed a back-up in case the first fell into the sea—put extraordinary pressure on Gup's builders. Requests became orders and orders became demands. Through what Gup called "a blistering array of memos," management decisions became Gup's concern.

As the office staff expanded, Gup also found the manager's demands intruding on the engineering aesthetic, those "measures of goodness, elegance, economy of form, perfection" that he'd prized as ideals for his team. It was unclear whether John's schedules were artificial restraints imposed for discipline, or whether they were mandated by necessity. Gup, as well as the other primary engineers, suspected, really, that John was no longer one of them. Instead of being protected from the stultifying conditions they imagined they'd escaped in the aeronautics industry, the engineers found themselves in a bureaucracy of John's own making. It felt like a betrayal. Gup began to suspect the worst: his builders were being used not to fulfill a shared dream but to further the project manager's career.

John resented these implications. He would hear their complaints through the grapevine, and grouse. "Hell," he'd say, "I have more engineering degrees than the rest of the engineering team put together, except for [Mark] Drela. I've been a part of critical engineering decisions all along. Have they forgotten that I spent hundreds of hours in the hangar with Chrysalis and Monarch and Emily?"

Still, it was true that he set artificial deadlines. Tight schedules had always played a role in their projects, and ever since he and Gup and Parky built Chrysalis, his strategy had been to calculate a reasonable amount of time for construction and then set the schedule slightly short of what was reasonable.

"The project's gotten to be a big operation," John said. "It's just that simple. The good thing about growing is we can do more creative engineering, but the bad thing is we're more hierarchical and require far more leadership. You can't have just one leader. I've been trying to set it up in three divisions, of engineering and flight operations and public relations. But I've had problems . . .

"Gup basically claims I'm never in town. And I claim he never shows up for meetings, and he stifles anything I do to communicate with the engineering team. And the project's lagging seriously. I'm disgusted, and he's disgusted. I mean, with the hacker thing it was okay to resist authority, but now, it really pisses me off."

One morning at Hanscom, conflicts between the old friends finally became insurmountable. John cornered Gup in the office and told him some officials from United Technologies wanted to tour the place the next day. The solar home was a mess, and it needed to be straightened up for the visit.

"Okay," Gup said, "that's good. Why don't you guys meet me out in the solar home and we'll start to set it up, because I'm not going to clean up by myself."

Juan wasn't around. Mark wasn't around. None of the UROPs had shown up for work yet. Gup strolled over to the office and asked again for help.

"Look, John, if you want this place cleaned up, why don't you come on out and we'll clean it up?" John didn't say anything.

Gup went back to the solar home to get started. After a while, Peggie dropped by to say she and John were going to lunch.

"Peggie," he said, "I thought we had to clean up for the sponsors. Aren't they coming tomorrow?"

"Yes," she said.

"Peggie, what does it look like I'm doing?"

"Oh," she said, "you're cleaning up."

"I didn't ask to do this job," Gup said. "Why don't you and John come on down here and help?"

After Peggie left, Gup worked a while longer and then he went to a phone and called the office. Peggie answered.

"John's busy right now," Peggie said.

"Look," Gup said, "he scheduled this meeting for tomorrow. I didn't ask him to do this. He said he'd help."

Peggie went out to the solar home and tried to explain again that she and John didn't have time and, besides, it was his responsibility to oversee construction. Gup exploded. He grabbed a broom and slung it.

"What kind of project is this!" he said. He pounded his fists on the table. "What kind of working together is that? It's not my job to clean up this place and yet he wants the place to be cleaned up for his meeting with UTC!"

Later that day, John and Gup confronted each other off in a side room at the hangar. They argued about schedules and engineering practices and the project's ideals, and then they came to terms.

"Who the hell are you to make these decisions?" Gup demanded.

"I'm the boss," John said. "I'm the manager."

Gup paused. "No, John, I don't work for you. I work for the project."

"I know you and Juan and Mark didn't sign up for me to be your boss," John said. "But there is a hierarchy here and I don't lord it over you and I don't push it off on you . . . and I know you didn't sign up under those terms, but we'll just have to agree to disagree. Mark and Juan don't make a big deal out of it. You just have to build the airplane. The rest of the staff and I just have to keep making decisions."

After that, their decade-long friendship disintegrated rapidly. Gup quit attending staff meetings. He stopped handing in his weekly time card, so despite continuing to work, he wasn't paid anymore. He, like Mark and Juan, wouldn't take the time to read memos in John's memo book.

John wanted to fire Guppy. But because of their history together and because there wasn't a finer craftsman on the team, he resisted the impulse. Instead, he promoted Dari Shalon.

Privately, John arranged for MIT to provide Dari Shalon full tuition for the fall of '87, a UROP salary of $360 a week, and a free place to live. Dari eagerly accepted the offer to co-direct the construction effort. More than ten years younger than Gup, Dari had precisely the kind of personality and desire John wanted to drive the young builders toward deadlines. Dari was a dreamer, like Gup, sometimes impractical and romantic, but he liked to bear down and he liked to manage. What some people didn't know about Dari was that during his years away from MIT, he'd not only traveled through Europe but he'd devoted a couple of years to military service. The effects of the training were pronounced.

"After being a commander in the Army you realize that managing people or leading people in civilian life is trivial," Dari said, later. "Literally trivial. Because in the Army you also have to worry about people's well-being. You worry about their mental state, their physical state, their food, their clothing, whether they are alert all the time. And the most important thing you worry about is them being motivated. The motto in the Army is, There is no such thing as a bad soldier; there are only bad officers."

Dari didn't ask for a title, and he didn't get one. Instead, Dari bought a used sports car and sped through the back roads from the project house in Lexington to Hanscom almost every day, squealing through stop signs, taking curves at sixty miles an hour, arriving home at 3:00 or 4:00 a.m. He invited the builders to share cappuccino with him. He read Tolstoy until dawn. He played Vivaldi on a tape machine in the hangar while the construction team worked. Dari was the epitome of dedication and discipline. The UROPs fell easily under his command.

As friction continued to divide John and Gup, a new alliance formed between Dari and John. The airplane construction moved ahead, and a long-standing bond between two former rocketmen dissolved.

One week, John was in Greece, confidently assuring ambassadors and Greek sponsors that he had the project under his complete control, and the next week he returned to Boston to discover that the corporate funding for Daedalus had become a tug-of-war among Shaklee, United Technologies, and MIT. John had often complained that he had never been integrated as a full member of the dealmaking process with Jack Kerrebrock, who negotiated with United Technologies, or Brian Duff, who handled the contractual arrangements with Shaklee. And he heard complaints from both companies' public relations offices when he hesitated to grant them exclusivity on any aspect of the project. But as far as John could determine, neither Duff nor Kerrebrock had bothered to coordinate their efforts, and the project fell under attack from all quarters.

When he returned from Athens in late September, John discovered that United Technologies' contribution to the project had increased to $470,000 and Shaklee's had been reduced from $150,000 to $60,000. "UTC essentially bought the Shaklee logo off the airplane," John said. "Kerrebrock gave UTC everything they could be given, so there was nothing left to offer other sponsors."

According to Kerrebrock, John simply didn't understand the original contractual arrangements. It had always been clear that in exchange for its support, United Technologies would have final authority over any markings on the airplane. "It's not a matter of allegiances," Kerrebrock said firmly. "It's a question of the normal way to do business. You make an agreement and you follow it. So when it came time to get money from Shaklee, I had to say it would be done in accordance with these rules."

A twisted kind of paradigm had arisen within the project, Kerrebrock argued, a kind of idealism that naturally conflicted with the commercial needs of sponsors. "Some members of the team just expect other people to give up when there's a hard point," Kerrebrock said, apparently referring to John's insistence about retaining control over the sponsors' commercial plans for Daedalus. "Whenever there seems to be any threat to the integrity of the project they expect everybody to fall over and play pussycat." Kerrebrock made it clear that, as MIT's representative, he'd never acquiesce to the Daedalus management team. He was, in fact, their boss.

When Shaklee's commercial possibilities faded, the health company's officials decided they still wanted to do business with Daedalus. So they approached Ethan Nadel directly. All along, Ethan had assumed he would have to prepare an exercise drink for the Daedalus flight, a job that could be done without formal research in the lab. But when Shaklee offered him $70,000 to conduct a full-blown research program—to quit acting as a project engineer and conduct verifiable scientific studies for the new drink—he seized the opportunity. "It was negotiated in not an ideal way," Ethan said, "but it came out that Shaklee divided up its contribution by giving me some money and some money to the project. Shaklee's motives were sound. They could help the project by getting a better drink, and do it without extra expenses. It wasn't altogether altruistic, but it was good business, nevertheless."

Tensions between Daedalus management, MIT, and the commercial sponsors intensified. John and Peggie exerted whatever authority they could to maintain a sense of integrity about the flight. But arguments with the public relations staffs at Shaklee and United Technologies became more frequent and even the university intruded on the spirit of the Daedalus flight. Petty fights over commercial plans, advertising, and public relations flared into furious disagreements over the meaning and significance of the project. Everything from its logo to the kinds of clothing the pilots would wear fomented prolonged debate.

An artist at the National Air and Space Museum had originally suggested a design for the team's emblem. The crew, as usual, modified and customized it, and the final symbol sprang from Juan's imagination. It was shaped like a disk, a circle surrounded by a border showing the angular, narrow chambers of a labyrinth; in the center of the circle was a drawing of Daedalus, leaping, his wings unfurled. A simple but aesthetically satisfying line drawing, Juan's mandala pleased team members and was accepted—oddly, without argument—as the official insignia. Juan's patch became known, affectionately, as "the meatball."

Naturally, the United Technologies staff abhorred the meatball. The sponsor's public relations team envisioned a logo with more pizazz, a sleek, modern design with bright red and blue colors that would shine through transparent Mylar and shimmer unmistakably from the pages of *Time* magazine.

While the corporation paid an internationally known graphics artist in Connecticut to create such a symbol, John, Juan, and Dari griped bitterly. The struggle over the symbol became a cause célèbre in the hangar, particularly after John found out how much money had been paid for a design that the team considered reprehensibly corporate.

The corporation's artist, Peter Good of Chester, Connecticut, was called into a meeting one day with a group of angry Daedalus engineers. Peter displayed his latest version—a sparkling modern figure, a kind of red and blue triangle that didn't outright say Daedalus, but suggested a mythic spirit, a symbol of flight at fin de siècle.

That's a dorky-looking thing, John thought. "That looks too nautical," one engineer said. "What does it have to do with Daedalus?" said another.

Peter explained that the red delta signified Daedalus.

"Well," John said, "if that's supposed to be a D, it's fallen over on its side."

The engineers heaped one insult on another—it was too corporate, it was too modern, it didn't make sense.

Peter Good listened politely. His wife, who was also an artist, finally interrupted and suggested to Peter that he might think of the engineers as artists, too. Peter listened to the rest of their criticisms and then returned to Chester to design another logo. He found working with the engineers a trial. "They seemed very dedicated and idealistic," Peter said later, "and I appreciated their craftsmanship. I had the highest admiration for them. But there was a conflict between their kind of philosophy and the philosophy of the corporation, which is more pragmatic, and which is a fact of our industrial world. . . . It became a major task, and it wasn't always pleasant."

The artist's final version of the Daedalus logo became known to the team as "the VISA card." The word "Daedalus" was spelled in Greek and English with blue and brown color bars shading the letters. The engineers continued haggling with United Technologies over whether they would really have to replace Juan's "meatball" with the corporate "VISA card." One of the engineers even suggested, cynically, that they block the logo by notifying the credit card company's lawyers to investigate trademark infringement.

A few weeks before the Daedalus airplane went on public display, United Technologies showed samples of its invitations for the rollout to John and Peggie. The VISA card logo appeared on the cover.

"This isn't the team logo," John told the corporation's public relations department.

In the end, when invitations arrived in mailboxes and offices, there was no emblem at all. No meatball, no VISA card. Nothing.

Taken as a whole, the myth of Daedalus spoke strongly to the point of conflict between engineers and kings. In Athens, in Crete, and in Sicily, Daedalus had been constantly in service of royalty. In every instance, in order to use his skills and "follow his bliss," as the classicist Joseph Campbell might have said, he had to provide entertainment for wealthy patrons or build them instruments of power. The same, of course, was true in the life of Leonardo. And in smaller ways, it was still true of countless engineers from Lockheed to Boeing to McDonnell Douglas. Aware of the "paradox," as John put it, he and Peggie worked to keep the project from becoming a mere entertainment for its benefactors. Engineering was not meant to be promoted as a spectacle on the Daedalus Project. It was serious, life-enhancing work.

"UTC came in to take over this little naive student project," Peggie said, "and they came in and said, 'Nice little nerds . . .' Pat, pat, pat. And they came in and found out we were not take-overable. We were not to be manipulated. We did not plug into the system. We did not plug into this thing that UTC thought we could be—this cute little group of engineers who could be used. We were a Daedalocentric universe. We had our image and we were strong-willed and intelligent individuals who weren't very grateful. And, it was like, 'Give us your money and get lost.' So they didn't get lost and we didn't act like we were supposed to."

The truth was, when the project first began, John hadn't attached any special significance to the myth. It was a convenient "hook" that would draw interest from sponsors and lend his team credibility. Over time, though, he'd read Mary Renault's novels about Minos and the Minotaur. On his weekends at the Cape or during airplane flights to visit sponsors, he read popular books about archaeology, which heightened his attachment to the Daedalus myth. The reality he came to perceive was not one that either his employers at MIT or his corporate benefactors could fully comprehend.

During the summer and fall of 1987, as Sarah Morris might have predicted, the new Daedalus myth did not seem to be about flight at all. The Daedalus Project was consumed by the problems of a king's inventor. At times, it certainly seemed like they had taken a flight to nowhere.

But out in the hangar, unaware of the political shenanigans and backbiting, a crew of UROPs worked continuously. They programmed the foam cutter to slice out shapes of electric guitars, drums, and a saxophone. Billing themselves as "the hottest new group in aviation history," a collection of UROPs performed during the Daedalus spar test. As Juan hung plastic water-filled bottles up and down the length of the long carbon tube, gauging the strength of Daedalus' black spine, they acted out a wild pantomime under the wings and in the shadows of a half-dozen experimental airplanes. While their managers and sponsors tussled to maintain control over what would become the public's perception of Daedalus, "Juan and the Cruz Missiles" wailed in the lonely Hanscom hangar, mimicking rock-'n'-roll musicians on mock instruments of pink foam.

18

Dari called for a death pact early in October. The UROPs skipped classes, missed exams, and lived on two or three hours of sleep for several days. More than two hundred dignitaries were expected at the hangar for the rollout presentation at the middle of the month, and the new Daedalus airplane still existed only as a mass of unconnected, uncompleted parts.

Late at night, like kids rummaging through a toy chest, the UROPs unloaded the solar homes and spread Daedalus out across the hangar floor in lengths of black carbon tubes, bits of shiny, milled aluminum, threads of golden Kevlar, and cleanly cut swatches of clear plastic Mylar, white Styrofoam ribs, and pink Foamular sheets. They fired heat guns, sawed with X-Acto knives, and dabbed glue with tongue depressors, trying to patch together a hopelessly undone airplane.

The day before rollout, October 26, three new pilots, Greg Zack, Erik Schmidt, and Kanellos Kanellopoulos, joined Glenn in the hangar. They were handed needles and thread and taken to a group of volunteers who were busily sewing the sponsors' cloth insignias onto steel gray team jackets. Once done, one of the pilots, Greg, grabbed a broom and mugged for a cameraman. "I can't do anything else, so I might as well clean up," he said. Minutes later, one of the engineers pulled Greg aside: "Come here, we're going to teach you how to build an airplane." Greg spent his first night with the project covering wings with Mylar.

Twelve hours before the rollout, one of the public relations officials from United Technologies dropped by to see the airplane for the first time. There was no airplane. John explained the old rocket hackers' theme—"Bureaucracy über Alles: The Impossible We Do Overnight, The Paperwork Takes

Forever"—but the man turned on his heels and headed straight for his car phone. "I guess he's never seen a death pact before," John said.

The Daedalus construction had always involved issues of aesthetics and measures of goodness. But these underlying values were the basis of their sacrifice, and Dari learned to talk about such matters as easily as Gup. They were like a whistle while he worked.

"It's all a matter of the standards we set back in the summer," Dari explained as they worked. "The tiniest things, even places you can't see, parts that will be invisible to the eye after it's assembled, will be beautiful. This is micro-design."

Working month after month, his builders had competed to make the most beautiful pieces, a goal achieved by using the least amount of glue for bonding, the smallest amount of microfill to smooth surface blemishes. Many times they had forgotten that those pieces would someday form an airplane. At the level of micro-design, they didn't see that the fuselage held a component part that fit into a gear box and turned the propeller, or that these thin tubes braced a stabilizer that would make an airplane fly straight. Twenty hours a day they had worked only to make clean chips of balsa wood and foam. "You get so absorbed sometimes in the detail that you forget the larger sight," Dari said. "But that's the way it should be. As far as I'm concerned, it's not an airplane. This is a work of art."

But as a result of their fastidiousness, the team lagged so far behind schedule that a fully operational airplane wasn't possible by rollout. A structural grid for the cockpit hadn't been completed, so they had to gerryrig a fiberglass support onto the inside of the cockpit to give it shape. The propeller wasn't done, so they mounted Emily's prop onto the spar with strapping tape. The drive train for turning the propeller still lay unfinished in the shop, so they decided just to leave it off and hope no one noticed. The construction effort, once overtly serious and slow, became an animated, theatrical affair. The sponsors had to see an airplane, even if it was just the shell of one.

The UROPs had built much of the airplane themselves, with only the guiding hand of veteran engineers. In fact, their own ingenuity led to the invention of tools unlike anything the old rocket hackers had seen before, a collection of handcrafted implements generically called "Izers." Tom Clancy's trailing edgerizer, an electronic cutter, sliced nearly three hundred feet of the wings' trailing edges and shaped Rohacell for parts of the fuselage in one tenth of the time it would have taken by hand. There was the carbon holerizer, built to cut holes in Juan and Claudia's paperthin carbon tubes on the fuselage. Dari's sanderizer perfected the holes at the center of the Styrofoam ribs so they hugged the spar tightly. Steve Darr, a UROP from Boston University, made the trimmerizer, a transparent plastic jig the size of a matchbox that allowed builders to trim the leading edge at its connecting joints up and down the wing. They used twisted devices called alignmentizers to join segments of the spar together. They created tools called threaderizers to make airplane parts fit together without a hitch.

One jig they avoided, but revered. Mark Drela had made a gleaming, hand-built motorized aluminum sander. Intended to be used for sanding balsa strips that wrapped around the outside of the spar, it turned out to be a spectacular kludge, an intricate, complex, bedazzling thing. It had been christened "The Rounderizer." Unfortunately, the machine didn't work. It whined so loudly and wobbled so wickedly that it threatened to eat the spar rather than sand the strips. The UROPs renamed it "The Randomizer," but after Dari invented the simple sanderizer, which worked perfectly, Mark's device was called "Anti-Christ." The UROPs wrote the name on a strip of masking tape and stuck it to the face of the machine. The Anti-Christ became a prized item around the shop, like the Finberg head, IDOLBERG. It stood as a symbol of the UROPs' increasing domination over the construction process—simplicity and elegance over sophistication and wizardry.

By 3:00 a.m. on the 27th, the hum of the hangar's artificial lights and the echo of rattling tools overtook the sound of human voices. In the glare of white lights, beneath an elephantine white canopy of aluminum and steel, the team worked steadily, quietly. A pungent drift of glues hung in the air. By dawn, Hanscom's mascot, a dog named Gyro, nosed around the airplane, sniffing among delicate parts without attracting a single comment.

The United Technologies official left his hotel room about eight o'clock and appeared at the hangar hoping to find the airplane in one piece. Daedalus still lay in piles. Gup had been up all night, and only a handful of builders remained at his side. The rollout was scheduled for ten o'clock. The poor guy stammered and stuttered, desperate for reassurance.

Two hours later, as visitors arrived for the ceremony, the Daedalus airplane hung just inches off the cold hangar floor, its wings spread out one hundred twelve feet, its plastic skin tight, clear, luminous. The fuselage was as white and smooth as eggshell; from prop tip to tail, the boom stretched twenty-nine feet, a clean black line of tightly wound carbon fibers. The eleven-foot propeller weighed 1.7 pounds. The entire airplane, lustrous pink wings and all, weighed only seventy pounds. The impossible, overnight. Gup was neatly dressed and alert. The UROPs wore ties and team jackets emblazoned with the sponsors' logos.

Tom Clancy's father, the sculptor from New York, approached the machine with awe. He'd recently completed a large piece of sculpture for Socrates Park just off the East River, made of a sharp steel blade that hung over a sheet of ice and a floor of concrete—an enormous piece, one hundred feet long, whose statement would be made over time, as the ice melted and the blade settled into the concrete and, inevitably, cracked the surface. In the past year he'd made other works (coincidentally called Monarch and Sparrow): large, temporal, modernistic sculptures whose beauty sprang from the ephemeral nature of the materials, the absolute permanence of their natural surroundings, and the poetic resonance of their names. Daedalus, he thought. What a piece!

The airplane was a work of art. You could walk under the wings to look at the internal structure—the Kevlar twine, the balsa strips, the lashings, every carbon tube laid bare—and see the imprint of human hands. You could see the unlacquered texture of sanded wood, the barely discernible shades of epoxy on carbon, glue on wood, microfill on Foamular. The transparency was exquisite. The wings and fuselage looked, to his eye, like stained-glass windows whose panes reveal the beauty of an internal structure.

Of course, there were the usual speeches from United Technologies, MIT, and the Greek government. When it was John's turn at the podium, he chose to tweak the sponsors. He'd wanted to entertain the assembly, but also to assert his team's strong and separate identity:

> Today, we introduce a new aircraft. If the hopes and aspirations placed on Light Eagle a year ago were heavy, those placed on this

new airplane must be crushing. Her task is to accomplish nothing less than taking the oldest human dream in aviation history and turning it into a reality. Today, we christen this aircraft Daedalus: thirty-five centuries of aeronautical aspirations are to be borne on a graphite backbone that weighs but thirty pounds.

The statistics about this aircraft contained in your program don't begin to tell the story: one hundred twelve feet in span, seventy pounds in weight, cruising speed of fifteen miles per hour. Behind all those numbers are people—a team of uniquely dedicated individuals. Today they are the pride of MIT. But over the years they have been the bane of more than one authority figure. These people compete to see who can take the longest time to complete an undergraduate education (the current record is seventeen years, but a new challenger is in the wings), who can open the most laboratory doors without a key, and even who can run the stop sign out front here at the highest speed (the record is 67.5 miles per hour, held by a faculty member). They have been forcibly ejected from dormitories, shops and even the Institute. Almost to a member, they hate paperwork and resent authority. In short, these are the kind of people that make MIT great. And they love this airplane.

We are often asked whether these particular airplanes have any practical value, and I for one almost always answer categorically no. That remains the answer today, but one that must be carefully couched in humility and uncertainty. Many of us on the team have come to believe that the technology developed in this project does have an immediate practical application, in terms of unmanned airplanes that would fly at high altitudes under solar power. Stationed for months or even years above a city, these platforms could provide services such as communications or observation. Anyone who has a few hundred thousand dollars and wants to bet on the next revolution in flight should speak to me after the ceremony.

We do not know where this plane will end up: we hope it will someday hang in the Gallery of Flight at the Air and Space Museum, alongside its cousin, Gossamer Condor. But we recognize that whoever takes off from Crete next spring will not be flying with any promises not also given the original Daedalus and his son Icarus, 3500 years ago. The flight is pushing the limits of human knowledge and capability, and everyone involved recognizes clearly that it may fail, at least in the sense of not reaching dry land. And yet, even this will not be a failure.

The benefit of projects like this lies in their value to the human spirit, and the idea that modern technology can be applied to ancient dreams. It is an illustration of how engineering can play a positive, constructive role in our society, and how science and art stem from common origins in Western culture. It shows how technological projects can draw together people from many disciplines and even from several nations. . . .

John covered all bets. He addressed the dream. He tweaked the kings. He pitched the future project he privately called Phase 4, the development of a solar-powered plane. And yet the speech didn't appeal to everyone. Gup didn't appreciate the references to the duration of academic careers. Juan wasn't sure he wanted to be a part of another business with John. Despite the rhetoric about aesthetics and art, there was, among the team, a sense of uncertainty and ambivalence when the ceremony ended.

After all, the next day the engineers knew they'd go back to work, refining the details of the first Daedalus airplane and cutting parts to build a second—just a precaution, a back-up, to cover their bets, too.

19

Women at the Sierra Pines apartments peeked out their windows when the cyclists rode home in the afternoons. They watched the pilots slip through the yard wearing black nylon Shaklee tights, their prominent muscles showing as clearly as if they'd just stepped out of a shower. The young men caught glimpses of ogling gazes as they walked up the stairs to their flat. Giggles greeted them; curtains snapped shut.

After only a few weeks in Lancaster, California, the Daedalus pilots were well known around the desert town, recognized in grocery stores and pharmacies. Bike shops offered them special deals on parts and physical therapists gave them free massages. Their training rides led them across Antelope Valley, through forests and wildlife sanctuaries, around lakes, into canyons. Even the Greek Olympian, Kanellos Kanellopoulos, had never seen such spectacular routes. They became, in a few weeks, as much a part of the desert landscape as condors and tumbleweeds.

But they were not happy in California. Heavy autumnal rains made it impossible to train on the lakebed at Edwards Air Force Base. On days when they could practice in Emily, John sent orders from Boston for them to join a small crew of Daedalus engineers for flight ops at the Mojave Airport; those who didn't fly had to run wing or stand around from four or five in the morning until ten or eleven. The athletes complained that management didn't understand their physical needs. They were lonely, too.

The pilots had been scheduled to train in California from late October to December, return to Boston for one week, and then leave for Athens with one of the two Daedalus airplanes. But after two rainy, windy months in California, they had gained little real flying experience. Steve Bussolari, assigned to manage the team, was constantly called back to MIT to cover academic obligations. The pilots grew gloomy and suspicious of one another. Erik Schmidt contracted a stomach virus. Glenn Tremml found himself working more often as the team's trainer than as a pilot. Kanellos Kanellopoulos talked regularly to his coach in Athens about returning to Greece. Greg Zack avoided his irritable roommates and spent most of his time flying gliders.

After two months living together on the edge of the desert, the pilots saw themselves more as lonely competitors than as members of a team. Occasionally, they tried to make friends. They signed their names to a Shaklee patch and stuck it to the front door. They taped a four- by three-foot map of Mojave on the living-room wall, and traced training routes through Antelope Valley into the Angeles National Forest. But they couldn't break the barriers, and over time they each staked out separate riding paths. Back in Boston, the Daedalus management worried over this new branch of potential renegades.

Inside the pilots' apartment, Glenn's bike rested against one wall, Erik's against another, Kanellos's against the third. There was a collection of dreary rental furniture: a nappy sofa, an armchair, a coffee table covered with cycling publications, girlie magazines, and parts catalogues. Shoes and pedals and bike tools littered the floor; boxes of new cycling clothes from Shaklee, sunglasses, and heart-rate monitors filled another corner. The ergometer and flight simulator commanded most of the room; a tawdry landscape painting hung just above the video screen. In the kitchen, the pilots covered the refrigerator with Chiquita banana and Dole pineapple stickers. They stored a case of Shaklee vitamins on top of the refrigerator. The continual chaos of the apartment indicated more than a locker-room mentality, though. It was territorial combat.

The pilots argued. They analyzed the motives of the management team back in Boston. They predicted who would be selected to go to Greece and who would get cut. At the Sierra Pines apartment, the pilots became acquainted first hand with the psychology of the Daedalus Project—for better, for worse. The pressures and uncertainties of the flight to nowhere weighed heavier with each rainy, miserable day.

* * *

At first, John thought Erik's quiet, intellectual nature recommended him as one of the best members of the new pilot team. He could discuss abstract expressionism and American radical politics as well as glucose metabolism and threshold phenomena. He'd impressed team members in interviews as the kind of person who'd work as a museum curator or spend his summers on archaeological digs. Small, thoughtful, intense, Erik's manner suited the image of the project.

Too bad they'd never seen him on a bike. Cycling was, in fact, Erik's greatest intellectual challenge—the mastery of mind over body—and when he trained, he rode in a rage. If John had known Erik better, he might not have hired someone so intent on becoming a great athlete. Unlike the other American pilots, Erik was on the brink of becoming an Olympian, and he didn't really need Daedalus to make the leap.

He'd been a product of the counterculture of the San Francisco Bay area. His father was an independent film maker, a producing director for PDQ Bach, a man who'd reared his son, at times, on a houseboat in Sausalito. Erik's knowledge of art and politics had come, naturally, from his father, but his interest in athletics stemmed from a need to distinguish himself in ways almost foreign to his father's generation. As a teenager he'd raced competitively on the West Coast skateboarding circuit; he'd surfed in college; he'd competed in triathalons with a zealousness that exceeded his natural capabilities. Erik pushed his body to breakdown for one purpose: to test limits.

In 1983, he'd placed thirtieth in the Hawaiian Iron Man Triathalon, an international event whose name conjures a competitive field of giants. But Erik—five feet eight inches tall, one hundred thirty-five pounds—had only entered the top ranks by risking his health. He wasn't a natural athlete but a gutsy competitor, the kind of cyclist who might vomit during the last leg of a marathon, and five seconds later find the resolve to surge ahead of the pack.

Since 1985, he'd raced in Colorado, New Mexico, Wyoming, throughout New England, Wisconsin, Nebraska . . . and he won. A string of successes led him to Boulder, Colorado, where he found a community of cyclists, and by 1987 he was touring and winning so many races that he was invited to race with the U.S. National Team. In only two years, Erik had made a meteoric rise on the amateur circuit. Cycling became a solitary passion.

"It's been sort of a heady time for me," he explained one afternoon at the apartment. "I never saw myself as a national-class athlete. I just saw myself as untalented Erik Schmidt who happens to get where he does by hard work. I rode very well during the last couple of years. You know, in some of these races I'd see myself passing all these top guys in the pro/am field, and it was just plain funny—after a race, I'd sit back and say, 'My God! What am I doing?' But I guess I've just been meeting my own expectations."

Before joining the project, he'd qualified for the national championships and then the Pan Am trials. When he read about the Daedalus Project in a cycling publication, though, the mythological quest piqued his interest. In the midst of racing season, he applied to be the Daedalus pilot; but the day after his interview at Hanscom, Erik was also invited to join the U.S. National Team for the Tour of Guatemala. It hadn't been an easy choice.

John and Peggie lived for weeks with Erik's initial ambivalence. His power scores exceeded all the Americans', and most members of the team were impressed by his interview. Not until after he'd moved to California did they discover that Erik was not as quiet or as pliable as he seemed.

Reports from Lancaster suggested Erik wasn't a "team player." Peggie asked him to make an effort to make friends with the engineers, but he told her that he'd get to know them in his own way, on his own time. He was naturally reserved around strangers, and he couldn't manufacture an affection for people he didn't know. "I'm really used to running my own life," Erik would say, "and being very independent. I'll go so far as to say that I'm used to being very self-centered and selfish with my racing career. You have to be to get to a certain level in athletics. And I'm not going to change that pattern."

John concluded that this particular engine posed a threat. Reports from Glenn, who roomed with Erik in California, only underscored John's worries: Glenn wrote that Greg and Erik were having problems getting along, and in one memo Glenn said that he did not consider Erik "altruistic." Anticipating that Erik would drop out or be fired before they ever left for Greece, John cut his salary from $2,000 to $1,500 a month—along with Greg's—to hire a fifth pilot as a back-up.

With his salary slashed and his odds for piloting Daedalus suddenly decreased to one in five, Erik grew increasingly sullen, disgusted with the project's management and uncertain of his future on the team. When he discovered that Glenn had been passing along "observations" about him to John, he grew suspicious and angry. Political jockeying with the management, it seemed, would eventually influence the decision about which one of them would make the flight. After only a couple of months on the project, Erik became an outspoken critic of the management.

Frank Scioscia knew Erik from the Pan Am races. In fact, he knew almost everybody on the cycling circuit. After twelve years as a competitive racer, Frank had established a regular path across North America, entering major cycling events from coast to coast. Frank also ran a side business, selling spare parts, cycling garments, and tires out of the back of a broken down '76 Cadillac. He was a wheeler-dealer as much as a racer, a gregarious, instinctual, itinerant ballyhooer. Characteristically, Frank won a spot on the Daedalus team in late November by charming the staff, sending a selection of Greek cookbooks to Hanscom and writing endearing notes to the office workers. Outgoing and carefree, Frank was a natural foil for Erik's intensity. He seemed the perfect back-up candidate.

Reared in Scranton, Pennsylvania, Frank was the son of a small businessman. His father, "Scratch," owned a new and used bookstore and a couple of parking lots, and liked to drive down to Atlantic City to play the odds. The father's talents were inherited by the son, who might have used them as a lawyer or a small-town politician eventually if he hadn't been smitten with a passion for racing bikes.

He'd been lured into cycling after he won the Pennsylvania state championships in 1978, only his second year as an amateur racer. By the time he turned eighteen he was invited to join the country's most promising cyclists at the Olympic Training Center. Although Frank had excellent grades in college, when he graduated from Lehigh University in 1983 he shucked job offers and ignored queries from graduate schools. He thought, quite innocently, that he could be an Olympian.

"Cycling always had to be subservient in my life to one thing or another," he said. "When I was a senior at Lehigh and started interviewing for jobs, I finally sat down and thought, What is it you really want to do? I could go to law school. I could go into business. But I decided, let's see how good I can be if I'm just doing bike racing and nothing else.

"A lot of people thought I was crazy. I mean, I wasn't so much of a talent that people said, 'Frank, you should give it up and go for the Olympic Team.' It was more like this: 'Frank, you're not that great; we don't understand why you're wasting your life on a bicycle.' "

He sold vacuum cleaners, aluminum siding, and AT&T phone services during the off season. If he had a commercial cycling sponsor, he'd take the bike clothes they'd given him and sell them at the end of the season. He'd start a new season every January with $4,000 in the bank, and then train steadily for three months. He made enough money to buy the Caddy and extended his racing territory beyond the East Coast. After three years he averaged one hundred races a year. He had a black address book filled with names of supporters, an informal circle of friends who would welcome him with a couch or sleeping bag whenever he came to town to race.

But it was a mean hustle. After five years on the road, Frank Scioscia still hadn't made a bid for the Olympic Team. In 1986, the coaches of the U.S. National Team listed his name among the top forty cyclists in the nation. But by 1987, when he joined the Daedalus Project, Frank's face had the care-worn look of a middle-aged man. "I've moved myself into the inner circle of the best people in the country," he said. "But it's taken me five years to do what I thought I could do in one. And in the last year, I've really floundered.

"I shouldn't even say this, because I've said I would eventually quit cycling. My immediate goal is to go to Greece with the Daedalus Project and be part of the effort. But when that ends I want to go full bore and make a mark at the Olympic Trials. Realistically, I'll ride in the trials, I won't make the team, and then I'll go to law school. I have thirteen applications in right now—from Harvard to Stanford to Berkeley and on down."

For Frank, Daedalus offered an opportunity to advance a flagging career. The project paid him a modest salary and gave him an inexpensive way to train for the Olympic Trials. The odds of making the Daedalus flight might not have been strong enough to bank on, but once Frank joined the team, he began to see other opportunities to be gained as a pilot—law school admission officers would look favorably on his involvement; maybe one of the project's sponsors would even want to endorse a cycling team for a year. The Daedalus management team didn't know it, but Frank Scioscia probably had the greatest potential for turning the flight to his own advantage of all the pilots. Frank understood the value of hidden opportunities.

The Greek puzzled everyone. He was as stoic as a stone and nearly unapproachable. At thirty, Kanellos Kanellopoulos had been a member of the Greek national cycling team for fifteen years. He'd spent a major part of his life in dormitories at the Olympic Training Center and at the National Academy of Physical Education in Athens. For thirteen straight years he'd won the Tour of Greece and ranked as the country's leading cyclist. He'd raced in the 1984 Olympics in Los Angeles, and earned a heavy cache of trophies from races throughout Europe and Africa. His publicity photos usually pictured him posed with a scarf around his neck, looking grim and jut-jawed like a Hollywood cowboy.

Once he arrived in the United States, Kanellos let his blond hair grow long and wild, and insisted on training alone. Some members of the engineering team thought Kanellos wasn't very smart, that he'd been reared for one purpose, and had never developed social skills. But that was just an American prejudice. Erik knew better: Kanellos had "the craziness" inside, an excessive need to race and to win.

Kanellos was born to a part-time olive farmer and house painter in the village of Vrachneika, near Patras in the northern Peloponnesus. He had picked olives and raisins and helped his father paint houses as a child. He'd dreamed of being a pilot in the Air Force. As a schoolboy he drew pictures of fighter planes in dog fights, men falling through the air in parachutes, and imaginary air raids from World War II. He had dreams of flying, too, which lingered even as he grew older. They weren't unusual dreams for a poor farm boy, but he was the only one among the Daedalus pilot team who had experienced these visions.

"I have seen my dreams like swimming in the air," he said. "Very often in my dreams I never fall. I fly safely. I float. I swim in the air. I think it is because I like to do different things and better things than other people do. Things that others cannot follow."

Kanellos Kanellopoulos

Glenn Tremml

Frank Scioscia

Erik Schmidt

(*Daedalus Project photos by Michael Smith*)

Greg Zack

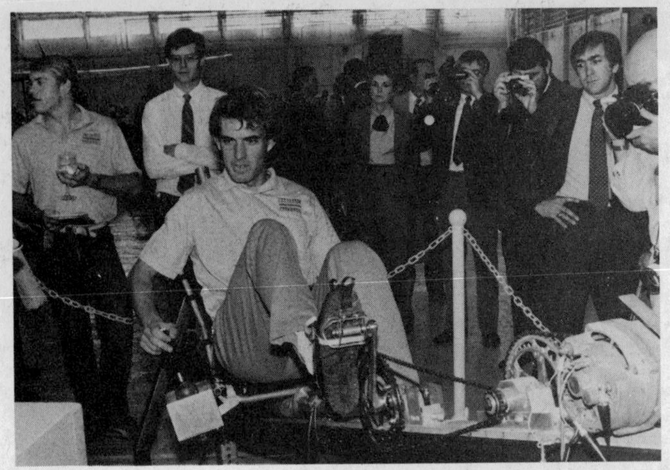

Kanellos demonstrates the Daedalus simulator after his introduction to the American press at the Daedalus rollout, October 1987. (*MIT Lincoln Laboratory*)

Daedalus at its rollout, October 1987. (*MIT Lincoln Laboratory*)

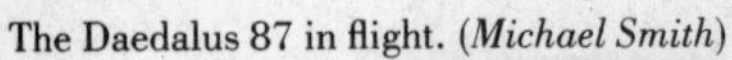

The Daedalus 87 in flight. (*Michael Smith*)

Frank Scioscia, John Langford, Eric Schmidt, and Steve Bussolari (*kneeling*) discuss pilot training in January 1988. (*Jim McHugh, Visages*)

By the beginning of 1988, team leader John Langford sometimes felt just like this picture: way out front and all alone. Langford and the Light Eagle, January 1988. (*Jim McHugh, Visages*)

Commander of the Greek Air Base (*left*), the Archbishop of Crete (*center*), and John Langford (*right*) during the Archbishop's visit to the Daedalus hangar on Easter Sunday, 1988. (*Chuck O'Rear*)

The Daedalus pilots at Knossos, on Crete. *Standing*: Erik Schmidt. *Seated*, R-L: Greg Zack, Glenn Tremml, Frank Scioscia, Kanellos Kanellopoulos. (*Chuck O'Rear*)

Daedalus 88 en route to Santorini, April 23, 1988. Marc, Dari Shalon, and Grant Schaffner man the inflatable below and wave away reporters. (*AP wire service photo, Michael Smith*)

Daedalus 88 soars en route to Santorini. (*Michael Smith*)

Daedalus in flight towards Santorini, April 23, 1988. (*Chuck O'Rear*)

"Aeolus slammed his fist down" — Daedalus 88 begins to come apart above the beach off Santorini. (*Chuck O'Rear*)

Daedalus 88 breaks up off Santorini, April 23, 1988. (*Chuck O'Rear*)

The triumphant Daedalus team on the beach at Santorini. (*Chuck O'Rear*)

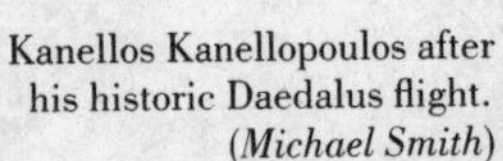

Kanellos Kanellopoulos after his historic Daedalus flight. (*Michael Smith*)

John Langford with the Light Eagle. (*Jim McHugh, Visages*)

Cycling had always been his secret means of escape. At thirteen, he quit eating lunch at school and hoarded his money to buy a racing bike. With the help of an uncle, he ordered a French bicycle for 4,300 drachmas, about $35. When the cycle arrived in a box from Athens, he left his school books with the uncle and rode through their village, out into the countryside—ten quick kilometers through the olive groves—and let the breezes dry his sweat on a slow ride home so no one would suspect him of mischief. "The excitement," he said, "was terrific."

He joined the village cycling team, and began to win races in competitions almost immediately. His reputation spread. He chased cars and buses on his bicycle, and, against the advice of older cyclists, he raced in the highest gears. Something wild grew in Kanellos. At sixteen, he won an invitation to live and train at the National Academy. Kanellos, it was clear, had the mark of a future Olympian.

Over the years, he'd performed well for his country. But he'd never scored at the Olympics. In 1980, the Soviet boycott barred him from the events. In 1984, he'd taken a tumble midway into the race in California and fell out of the pack. By 1988, he had one more chance to compete in the Olympics, this time in Korea. Just as his training season began, he learned, instead, that he'd spend a good part of the winter in the Mojave Desert with a group of American amateurs. At the behest of the Ministry of Sports, Kanellos had been traded to the Daedalus Project.

He was philosophic about his predicament. Deals struck at governmental echelons far beyond his purview made it difficult for him to control or even understand the arrangements that had been made with the Daedalus Project management. He'd been told by his trainer that he would spend a month or two in America, training for the flight, and return to Athens in time to compete in the Tour of Greece. If he placed well in the event, he might go to the Olympics in September. In any case, in 1988, Kanellos knew he had a chance to retire either as an Olympian or, if he was chosen to make the Daedalus flight, a legend. The odds weren't promising—five to one to make the Daedalus flight; more or less the same for the Olympics, depending on whether the Greeks decided to send a cycling team at all. Either way, Kanellos sensed the end of his career in sight.

He rarely discussed his plans with the other pilots. His weekly rides through the mountains above Mojave stretched from four to six hundred miles, and when he took time off to relax around the apartment, he existed in such a solitary state that no one dared enter.

Despite prior arrangements, John wanted to keep Kanellos on the team through the Tour of Greece. The team's meteorologist predicted that Aegean weather presented a six- to eight-week flight window for Daedalus between mid-March and early May. How could he afford to lose his strongest pilot for two weeks in March? The Greek was a physiological wonder John just couldn't let slip away.

Disagreements flared between John and Kanellos's trainer a few weeks after the pilots arrived in Lancaster. In phone conversations and telexed messages, the trainer demanded that Kanellos be returned to train for the Tour of Greece, but John fought back. "Possession," he'd say, "is nine tenths of the law." Under orders from the Daedalus management in Boston, the project's representatives in Athens lobbied the highest branches of the ministries of sports, culture, and information to soothe Greek tempers. As the politicking wore on, the American pilots watched Kanellos relentlessly increase his training through the Southern California mountains. The Greek approached levels of duration and intensity that seemed, frankly, inhuman.

Of all the pilots, Greg Zack was the only one with a genetic advantage. The VO_2 Max tests and other physiological measurements showed that his body produced the same amount of power to fly the airplane as a comparable athlete—but he used far less energy to do it. Greg was the team's most efficient engine.

Unlike the others, though, Greg was almost finished with cycling. Five years as an amateur rider had taken him into the upper ranks of the sport. If he had ever dreamed of one day being an Olympian, by 1987 he found himself in a field crowded with strong, radical cyclists like Erik Schmidt and Frank Scioscia, who had more desire and a greater tolerance for pain than he ever would.

Greg was a local hero in Lexington, Kentucky, a part-time mechanic at Everybody's Bike Shop, a veteran of the Mid-Western cycling circuit and plainly an athlete by default. He seemed to be the kind of guy who'd never quite figured out what he wanted to do with his life—so while he was deciding, he raced bikes. In college he had studied the cello, then economics, then business. For years, the only constant was cycling. When it came time to graduate from the University of Kentucky in 1984, he had sampled such a variety of courses that he could have claimed a number of disciplines for his degree. Eventually, he did graduate, but he also kept racing. When the Daedalus Project presented itself Greg was waiting for a change. He needed an adventure.

Some people on the project thought of him as just a good old, skinny southern boy, with a wisp of sandy blond hair, an easy grin, and a silly sense of humor. ("As a Daedalus pilot, do you have to watch what you eat?" a reporter asked. "Sure do!" Greg exclaimed. "I watch it all the way to my mouth!") He certainly wasn't as savvy as Erik, Frank, or Kanellos. But in some ways he was the kind of pilot the project needed—his personal goals hadn't been defined yet; he had a deeper, purer, perhaps more naive interest in the values of the project. Like Glenn, Greg rarely questioned the management. He thought the project was a wonder, and he was privileged to serve it.

"The biggest shot to me was seeing the plane for the first time," he said, recalling the October rollout at Hanscom, when he met the Daedalus team. "It was all in parts, but the pieces and the attention to detail were so incredible. I would look at this huge wing—it was magnificent! And then I would come up under it and see the smallest pieces that held it together. There was no sloppiness at all. I remember helping the engineers cover the wings with Mylar, and they wouldn't quit working if there was a wrinkle in it. No way! It had to be perfect.

"Then looking at the Light Eagle, it was just as amazing. It was weathered and tough and beat up. It had battle scars. You could see where people had been sweating in it. Your imagination starts soaring, and you think about all the things this plane has done. I mean, I knew nothing about flying so it was totally new to me. Everything was incredible.

"When we got out here, that's when I finally got a chance to really learn how to fly. We started taking glider lessons out here at Cal City, and it just blew my mind. They have this airport in the desert, and the wind blows perpendicular off the mountain range, and it comes around following the contours of the landscape, bounces up, and circulates through the sky in waves. Under the curl the wind is very turbulent, and it makes a very interesting place to learn how to fly.

"One day I'm up there wearing a parachute, and my trainer's saying, this is kind of boring, huh? And I'm saying, you've gotta be kidding. We're at ten thousand feet and he says to me, tighten your belt! And then he says, tighten it some more! So I do that. And he says, mind if I fly? And I say, sure, and he goes sshhwwsshh, and all of a sudden I'm upside down and I'm like, what are we doing? And we're flying upside down. And he says, look at the variometer, we're still climbing. And I'm thinking, boy, this is better than anything I have ever imagined! I'm thinking, nothing's this good! All my blood's up in my head and my eyeballs are bugged out.

"Then all of a sudden he puts it nose down and we're doing this loop. The sand in the cockpit is scratching around the canopy and we're screaming around again. And by now it's, like, Greg, this is it! You do anything else in your life and it can't be any better than this!"

The other pilots always suspected Greg spent more time flying gliders than he did on the road pedaling. They were suspicious of his enthusiasm. Reporters enjoyed his gregarious nature, and quoted his home-spun humor. When orders came from Boston to attend flight ops, Greg went without complaint. But as the pilots started to log their weekly cycling mileage, at John's request, to send back to Boston, they couldn't believe Greg's high tallies. Erik and Frank, particularly, wondered whether Greg wasn't inflating his numbers. Was he trying to use the project to get a commercial pilot's license? Was he really a genetically superior athlete? Why did he pander so much to the management? What were his secret ambitions?

Greg was the only true believer among skeptics. He made them edgy and they made him grumpy. And it just made life at the Sierra Pines apartment a little bit tougher.

John had always said the project lived and died according to unknown unknowns—"unk-unks," that is. "An airfoil is merely an unknown," he'd explain. "That is, predictably a problem. The same with carbon tubes and propellers and Kevlar lashings. You prepare against Murphy's Law by testing them to failure, watching them closely in trials, exerting stresses and strains to see how they behave. On the Daedalus Project—except for the quirky Mediterranean weather—the pilots have the greatest potential for being unk-unks. Athletes can't be tested like carbon tubes, and their motives aren't always clear."

During the fall, John decided that Ethan and Bussolari should establish a rigorous training schedule and a set of hurdles—testing criteria—for the pilots. He continued to insist that the pilots show up for flight operations so they'd gain an appreciation for the airplane and its makers. He reviewed Glenn's weekly records of the pilots' training logs, and decided to leave the central issues ambiguous—how many pilots would serve in the final rotation and precisely what criteria he'd use for selecting the flying team. John told them he honestly didn't know how they'd be picked. There was always the possibility that one or two of them wouldn't go to Greece.

The athletes felt cheated. The competition among them was just too raw not to know how they'd be chosen. Would he take three of them? Four? Would someone ultimately become Daedalus by a high score on a VO_2 Max test? By piloting ability? By strength of personality or political savvy? What about the back-up?

The sudden conflicts in California alarmed the Daedalus management. Finally, in November, Peggie moved out to Lancaster to attend to the pilots' needs. She carried a couple of Macintosh computers, an electronic mail system, a laser printer, boxes of press kits, logos, and Daedalus T-shirts. She faced a maze of trouble in addition to the pilots' tempers: delays caused by the rains continued to push the flight training program deeper into the new year; sponsors complained that their companies' names weren't showing up in news stories about the project; Steve Bussolari had effectively abandoned the pilot training program and Glenn was again prepared to quit.

In December, after confirming a deal in Greece with Olympic Airlines, John returned to the United States and moved out to Lancaster, too. He took an apartment at Sierra Pines and, for the first time, got to know his athletes personally.

20

Glenn sat up at night gloomily doing calculations. It was a week before Christmas, and he'd finally encouraged the other pilots to act. He would help them plot a strategy and draw up a list of grievances to take to the apartment across the yard.

After three weeks together, neither the pilots nor the management had come to terms with the physiological demands of the Daedalus flight. Steve Bussolari had all but abandoned the athletes—he and his wife had a new baby to tend back in Boston. The team was surviving the rainiest season in California in fifty years, and the pilots had reached their limit living in a cramped apartment, waiting to fly, fighting off colds.

The numbers looked familiar. Glenn had to prove that one of them could, very likely, end up dog-paddling in the Aegean. He'd made the same argument to John and the engineers for six months, but no one ever bothered to listen. Now, if the pilots could just transcend their personal quarrels for one night and approach John and Peggie with solidarity and scientific evidence, maybe they could be persuaded to take the engines' problems seriously.

He ripped through the calculations one last time, and then moved into the kitchen where the other pilots argued over supper.

"We cut the training many times and we reduce everything," Kanellos complained. He stood in a corner chopping lettuce with a long knife. "The whole schedule is set back. They go like this: 'Fly the Eagle . . . No, not today, it is too wet . . . We have no car to take you to the airport . . . Go here—go there.' We spend a lot of time running around. Once, twice, it happens, okay. But three times, more, you get tired. You say, I don't care."

"Let me show you something, Kanellos," Glenn said. He snapped on the calculator and sat down at the kitchen table next to Erik.

"We know that ever since John came back from Greece, they've been leaning strongly toward a flight to Santorini rather than the mainland. That's to our benefit. All the weather experts have told them to forget about the longer route because of the landmasses along the way—too many unexpected winds and turbulence off the cliffs around Maleme. Anyway, they're beginning to buy that. So let's just look at the flight from Iraklion to Santorini."

Frank and Greg spooned up plates of rice and lentils and came over to the table.

"Santorini will be a seventy-four-mile flight," Glenn said. "Okay? And since the plane flies at fifteen miles an hour"—he punched the buttons—"seventy-four divided by fifteen equals four point nine. Four hours and fifty-six minutes."

"But that's assuming we fly a perfectly straight line with no wind," Erik said.

"Right," Glenn said. "So we have to consider that the prevailing wind in that region is from the north. And since Santorini is north of Iraklion, that means we'll likely fly into a head wind. Let's just assume the whole flight is within bounds—they've built the plane to take a three-knot head wind. So a three-knot head wind will lower our ground speed to twelve knots. Divide seventy-four by twelve knots"—Glenn's fingers flicked at the machine assuredly—"equals six point one six. Six hours, ten minutes."

Kanellos lifted his knife from the cutting board. "Four hours, it's okay. If you are in perfect shape. But six? Impossible! Impossibly impossible. That is definite. I can sign."

They all knew Kanellos could cycle eight or ten hours. Erik looked particularly alarmed.

"When I first heard about this project they said the flight will take four hours," Kanellos continued. "Then later I heard, maybe five hours. Now you say six hours! What is the truth?"

"I just don't understand," Erik said.

"I think," Frank said, "they just let this part of the project slip because the first person they had to fly the airplane—that's Glenn—was a pilot and an athlete. So now they figure they've got a higher breed of athlete. But in a way, Glenn set a bad example by already knowing how to fly and being so easily trained and fitting into the program. Now they're upping the ante physically, but they haven't taken into account that we've still got to learn how to be pilots."

"It's all guesswork, really," Glenn said. "It may not take six hours. We could get a very unusual tail wind and get there in three. Who knows, maybe Finberg's autopilot could work and we'd fly at night."

Frank laughed. He'd been living with Finberg and his crew of UROPs in an apartment downstairs for almost a month. They worked every night on the vaunted autopilot, and it hadn't come to life yet.

"They've just been pushing all the edges," Frank said. "Initially, this was an engineering project, and they've succeeded along those lines. But no one's made them think about the rest of it. I mean, it's luck to have six hours of perfectly calm air over the Aegean. Now it's just going to have to come home to roost. They have an appreciation for the engineering thing, but they don't have an appreciation for what it will take if you have to be in that plane for six hours."

As they ate—enormous salads, heaps of cooked grains, whole loaves of breads—they complained about Bussolari and Peggie and John. Erik griped about his pay cut. He wondered how a million-dollar project could be strapped for funds at the moment when they desperately needed a pilot trainer. Erik looked at Glenn. John had appointed Glenn to track the pilots' spending.

"What do you think?" he asked.

"About what?" Glenn said.

"If they say they can't afford a trainer?"

Glenn stammered. "I don't really know how it works . . . uh . . ." Did Erik know Glenn's salary hadn't been cut? "I'd just say that after you're involved in the project for a long time, you realize how it works. If you're taken on at the last minute you don't really know how it works and sometimes there are surprises. Like, there's a difference between the money we expect to raise and the money we have. And the salaries we expect to pay and salaries we can pay. And, you know, agreements that are really proposals that don't necessarily work out the way they were proposed."

"It's just like anything else," Frank said. "You learn the ropes after you get in it. You find out later what you have to do to work the system."

Erik shook his head. Nonsense.

"I'm not a volunteer," he said. "This isn't my dream after three years. I don't want to spend all my time building an airplane and delaying a career or whatever. I look at it as a job—one that's very interesting and one that I care about—but I don't like it when people try to change the rules after the game's started."

Kanellos's face flushed. His odds had suddenly changed, too.

"The Tour of Greece is the most important race in my country and is on the day of this project," he said. "But the officials of this project did not discuss it with the most important officials of the Cycling Federation. They talk with other people. My trainer is mad. At first they tell him, 'Sometime in early February the pilots will be in Greece. Kanellos will race.' Now they say it will be the last days of February. I cannot be ready, five months here, training away from my team, my coach . . . my country!"

After supper they agreed to speak with one voice. The major issues were these: The pilots needed a trainer; they should no longer be expected to attend early morning flight operations unless they were flying; and the possible duration of the Aegean flight, up to six hours, must be acknowledged.

At seven o'clock, with their bellies full and their ranks mustered, the five athletes left the apartment. They slipped through the dark and drizzle to the management office. John and Peggie had been expecting their call.

The project had grown too large for Peggie Scott's liking. Particularly since she'd become, over the past eighteen months, the team's press liaison, the financial officer, the secretary, the assistant project director, and the sponsors' punching bag. She was a strong, smart, aggressive woman, and she hadn't come into the project to do paperwork and field personnel complaints. It was art and engineering that drew her, the delight of building wings with a small crew—like she'd done on Emily—that held her interest. By December 1987, without any indication that she'd get more clerical help and increasing evidence that she'd never return to the building crew, Peggie felt discouraged. She'd taken jobs that no one else wanted or could manage. She was very weary.

When she opened the door, it didn't surprise her to greet five more disgruntled people. The athletes stepped gingerly over her boxes of public relations materials, press kits, and memos. She invited them to sit down, and asked if they wanted to have dinner with her and John. Glenn said they really just wanted to talk. Just as she'd expected.

Greg's head bobbed as he mustered the nerve to speak. He'd agreed to support his teammates, though it would be hard for him to confront the management. "It would be nice if all the engineers got on the simulator and tried to ride it at flight power," he said. "It would give everyone a better idea of what we have to do."

"Don't underestimate these engineers," Peggie answered. "Most of these guys can fly that plane better than you guys will ever be able to fly it!"

"Well," Greg said, "a two-minute flight or a five-minute flight. But I'm talking about a five-hour flight. I always get the feeling that everyone thinks that building the plane was immensely more difficult . . ."

"No!" Peggie said.

". . . than flying it."

"Well," Peggie said, "have you ever tried building a spar?" John let her talk. He, too, was tired of their whining. The time for diplomacy had passed.

"Well," Greg said, "I've stayed up until four in the morning as people tried to put the airplane together. I just guess the idea is that everyone thinks our job is real easy and . . ."

"That's right," Frank said. "As well as you're pushing the technology of the plane, you're pushing people's physical nature, too."

Peggie looked at John. Maybe she should try another tack.

"Okay, well, would you guys be willing to set up a seminar series about training and endurance exercise?" she said. "I don't know anything about your training. Maybe it would help us understand why it is you do what you do."

Erik, who sat on the floor, shifted up against the sofa and pressed the issue again.

"The point is that our need to train right now is crucial. And these demands that we show up for flight ops are detracting from our fitness."

"It would be perfect if I could ride the simulator every week," Greg said. "If someone would just watch me and tell me what I'm doing wrong."

Peggie looked startled. "So what you need now is somebody to watch you? How many hours a week?"

Frank turned to Glenn. "What do you suggest?"

"Two hours," Glenn said.

She laughed, incredulously. "So you want somebody to sit in your room and watch you fly the simulator for ten hours and give you instruction?"

"It doesn't have to be full time," Glenn said. "Just a few minutes. Bryan Allen [the Albatross pilot] had a trainer with him every single minute, screaming in his ear telling him what to do. Pushing him through it."

That's true, John thought. He looked at Peggie. "Well, I guess this is a good time to bring it up," he said.

Frank carried on with the negotiations. "See, if we start thinking that we're looking at a six-hour flight, then we have to consider, well, what's it going to take for us? I have a handle on the physical training. But I don't have a handle on how much simulator time or flight instruction it takes. I'd more than welcome taking guidance in that direction. But it seems that you guys set certain priorities for us, and then everything else is abandoned.

"I mean, today is just a perfect example. We had gliders reserved for two o'clock, and Greg and I were on our bikes training, and one of the engineers was supposed to meet us and drive us to the glider port. But nobody came by to pick us up until three-thirty. So we were just sitting out in Mojave waiting. It cuts into our time. And on top of that, once the engineer showed up, he'd forgotten to have any clothes for us to change into. So we couldn't fly. Now that's follow-through on stuff that someone should handle."

"What do you need?" Peggie asked. "A personal car?"

"No, it's just if somebody says they'll do something . . ."

"Peggie," John said, "I think what they're saying is we have someone from Shaklee to help them design their diets, and we have coaches for their glider lessons, and what they need now is a trainer to hang around the car." His sarcasm went unremarked. John was angry.

"Do you guys have any ideas?" he said. "You know what the universe is in terms of human-powered airplane people. We tried to hire Bryan Allen to be a full-time instructor and trainer. But he's just taken a new job."

"Maybe we should try to see if he's still in that job," Peggie said.

"Yeah, maybe we should," John said. "Any other suggestions? The floor is open."

"Now, could this position be for a simulator instructor?" Frank asked.

"It sounds like you have a bunch of things you need," John said. "Simulator instructor. Making sure that there's somebody there to help you get your clothes out of the damn van. I mean, I don't know. What do you think you need?"

"It would be great to have a team manager," Frank said.

"We can't be that organized," Peggie said.

"I know it sounds like the most asinine thing to you guys but like I was saying to Greg today, we shouldn't be *acting* like prima donnas, but we should be treated like prima donnas. In the sense that, like, I don't want to come in with a big head, but I know that this will make a real difference. But when you're talking about this kind of physical challenge, we have to ask for a lot of help."

Glenn could read the managers' faces. They'd pushed their luck.

"I just want to say that I understand completely what your problems are, okay?" Peggie said. "And I sympathize. But I think you're in fantasyland thinking we're going to be able to find someone immediately. And just as we desperately need a pilot who can pedal this plane across the Aegean, we also need a dozen other things from team members, who also need back-up support. We have limited funding. We have limited time. Bottom line—we'll find you someone, but you have to think about where payroll comes from."

There was a long, stultifying pause.

"I think we're almost done," John said. "It's a good time to do this. But it would be a mistake to try to resolve it now. We'll have a Christmas break to think."

Frank took a breath. "I'm sure everybody is behind the eight-ball. But if you could bring in someone else to work on the airplane instead of having me running wing, that's more important from my end. It's not that I don't want to run wing. But we've got to get down to what we're supposed to be doing because time's running out."

"That's right," John said. "The next seventy-three days will be pretty strenuous. Anything else?"

As the pilots walked back to their apartment, Erik turned to Glenn.

"I think Peggie was trying to hint that the salary for this hypothetical person will come out of our paychecks," he said. "Do you think she was trying to hint at that?"

Kanellos walked silently with Greg and Frank. "Peggie keeps asking about our VO_2 Max scores," Frank observed. "I think she's just trying to establish in our minds that we've got this extra reserve."

Greg shook his head. "If they'd just sit in the flight simulator for two minutes," he muttered.

Back at the office, the managers discussed their predicament. They couldn't afford to hire a trainer. Maybe, John suggested, they could take Lois off the public relations staff in Boston and bring her out to California to run the pilots' affairs. But of course, the engineering crew at Hanscom still had weeks before they'd finish building the second Daedalus airplane, and they needed some support staff there, too. The sponsors had lined up a series of promotional tours for the engineers and pilots over the next several weeks, and Lois might be needed to fill in there as well. If only the rain would let up.

Peggie was suddenly quiet.

"I don't have anyone running my errands," she said.

"I know," John said. He thumbed through his wallet to make sure he had money for dinner, then he paused. "We're just not ready, are we?"

"We're a long way from being ready," she said.

The next night one of the engineers' girlfriends organized a Christmas party for the team. Tom Clancy prepared a banquet of Mexican foods, fresh fruits, and chocolate chip cookies. They drank Corona beer. Greg ate a couple of pounds of burritos. Erik hung Shaklee logos on the Christmas tree. Glenn left the party early. He was depressed because he couldn't find a flight home for Christmas Day. His stomach ached. Finberg complained that his engineers were suffering from "data collection sickness." The autopilot was a disaster.

It was an unusually solemn Daedalus gathering. The engineers knew that until the weather improved, they would drift further and further from the spring flight window. They had billed their quaint get-together "The Christmas Wake." The hours were appropriate. The party lasted from 7 to 10.

A few days later, even Peggie decided she'd had enough. She typed up her resignation, packed her bags and left Lancaster for good.

Peggie had been John's most invaluable colleague. She'd alienated some members of the team because she could be tough and demanding, but she'd also devoted enormous amounts of time handling the deluge of paperwork, organizing public relations events, running the office with energy and enthusiasm. She'd risked friendships in the endless struggle to maintain the project's identity against the wishes of MIT and the corporate benefactors. But she was fed up and exhausted.

Finally, at his parents' home in Atlanta for the holidays, John sat down with a calculator and punched figures into one of Parky's old human endurance equations. He had always been certain that his pilots had plenty of margin—even in a head wind the flight couldn't extend much beyond five hours, he thought. But when the calculator showed the same figures as Glenn's—six hours, ten minutes—he erased the screen and refigured. Six hours, ten minutes.

There was no one left to talk to. He and his wife were barely speaking—the project had staked a wedge into their marriage. His friendship with Gup was over. John was no longer sure where Parky's allegiance lay. Even Mark and Juan had put some distance between themselves and the management team. In fact, John realized that the only person he really could talk to was Peggie.

With a mass of schedules and memos and contracts still left to draft, John celebrated Christmas with his family in Atlanta. The power plays and struggles that divided his team would continue into the New Year. And for the first time, John realized that he was alone.

21

How did he cope? An inner voice, denial, the darker qualities of an adventurer led John deeper into the enchantment.

This is not the way it's supposed to be, Barbara thought. She and John had been together for more than twelve years. His ideals and optimism had always girded their relationship. The corny way he quoted President Kennedy's speeches amused her. The airplane projects—from Chrysalis to Emily—had brought into their marriage a range of fascinating people. Over the years Barbara had spied on Paul MacCready, built pieces of Monarch, and snapped keepsake photographs of human-powered airplanes. But she had never expected Daedalus to loom so large in their marriage.

Her husband's plans began to seem foreign to her, deceptive, in the way that dreams can be deceptive. The compendia of charts, maps, schedules, and memos that piled up in his office at home *looked* like the concerns of a practical, responsible adult, but what was Daedalus, other than a big model airplane? The room was wired for computer linkages to California, Boston, and Athens. She watched him photocopy almost a million dollars in contracts there, and schedule flights that took him four times to Greece, regularly to Boston and Los Angeles, Florida, New York.

For the better part of a year, John was rarely at home. When he spoke of the myth, Barbara's eyes glazed over. If he spoke of the project's goals—increasing cultural awareness, experimenting with interdisciplinary research—she felt a twining knot of skepticism and anger. "Give me a break," she'd say. "All that's fine, but I don't see how this will solve any of the world's problems." In pursuit of dreams, John had left Barbara alone for too long, living in limbo, raising Ellis mostly by herself. Deadlines came and went. But the projects never ended.

In December John had taken Barbara to Greece, hoping to rekindle their relationship. He introduced her at meetings, took her to the Parthenon, tried to interest her in the details of his work. He thought she still seemed ill at ease, still threatened somehow. When they traveled to Iraklion, John experienced a strange sensation, "as if a force were reaching across the centuries and saying, 'This is where the dream began. This is where you will begin.' " Barbara didn't know exactly what he meant. "Dreams?" she said. "That's a very slippery concept."

At thirty, John Langford was learning, for the first time, about dreams. He felt strained by the demands of teammates and conflicts with sponsors; he felt guilty about leaving his family. He wondered what he had given up and what he had gained.

Like celebrities, politicians, and million-dollar managers, he did much of his thinking aboard airplanes, where there were no horizons. The orchestration of Daedalus felt wonderful when he rode in the air. His briefcase contained correspondence with film makers and writers who wanted to tell the Daedalus story. He prepared presentations for investors interested in the solar-powered airplane project he wanted to develop next. On these flights, he could even envision sending an unmanned plane to Mars someday, commanding a crew of professional engineers.

During the first days of 1988, after the holidays, he flew to New York for meetings with reporters and editors at *Time*, CBS, and the Associated Press. In only three months the Daedalus team had to be in Greece. He slipped on his gray flight jacket over a black pin-striped suit and lifted a large cardboard box and a hard-shell aluminum briefcase. He looked like a barker for a flying circus as he hailed cabs around Manhattan—one part engineer, one part Wall Street. In the executive suites of Madison Avenue offices, he whipped off his logo-filled jacket, unstrapped his cardboard box, and pieced together Gup's toy model of the Daedalus, talking a patter about aerodynamics and mythology and teamwork. The wing span of the model stretched across entire offices—New Yorkers smiled and murmured.

And the tie? they'd ask.

Oh, these? he'd say, looking down at a thin field of tiny human-powered airplanes swarming like bees. That's Monarch. We had these made after the world speed prize.

And the pin?

Daedalus, he said. People on the team call it the meatball. It's a . . . personal emblem.

He never failed to mention the names of Parky and Guppy, as if they were all still close friends, just out of the rocket society, and Bussolari and Ethan and each of the pilots, as if they remained a part of the unbroken circle, too. He seemed to forget that his friendship with Gup had dissolved, that the pilots often saw him as a scourge in their lives, that his "team," like his wife, wondered what had happened to him.

While he was away, the engineers griped about the changes that characterized John during the past year. None of them showed much interest in "Phase 4," the hypothetical solar-powered airplane business. "Is there life after Daedalus?" Juan would say. "No! There is only Phase Four!" The engineers would laugh derisively. Even Parky, who was working eight hours a night in his shop in San Jose to build the new Daedalus gearboxes, had reservations about joining John again on a start-up venture. John has removed himself from engineering, Parky observed. Is this project about engineering or is it an ad-venture? Another lead engineer complained: "He thinks he can solve all problems with an organizational chart. He needs a Macintosh and an organizational chart. Once it comes out of the Mac, it's supposed to be true and finished. Problem is, people don't always fit on an organizational chart."

Pursuing the dream exacted its price. Over lunch with a literary agent, he looked like a cardboard man, two-dimensional and pale. It was almost as if he'd arranged these meetings to reassure himself that all was well. Of course, he'd never unburdened himself to a single member of the team. Strictures of leadership imposed solitude. He had to be decisive, optimistic, firm, fair. Only once that winter had John shared his problems with anyone—one night in Boston he met a former MIT classmate and they drank a few beers, traded woes, and parted. Otherwise, no one knew. The sponsors received enthusiastic letters from him; a small group of investors were courted for funding the solar-powered airplane project; the literary agent heard the charming story of the project's origins. John wore resolve like a shield.

As his teammates grew cynical about these affairs, John seemed to escape by living in transit. United Technologies' public relations appearances became a kind of a blessing. By contract, the project was required to set aside eight days a month for public appearances—he honored at least half the corporate requests. No one else really wanted to stand and speak on the dream's behalf.

He also burrowed into a document he called the Operations Plan. John practically lived according to the paper trail of the Ops Plan. Over months he had distilled every aspect of the flight in his Macintosh—the chronology of readiness, project logistics, the timing of the boat flotilla, communication relays across the Aegean, abort contingencies, support teams, back-up support teams. He and Bussolari had drafted the first Ops Plan on a flight out of Athens during the previous June, and John spent an increasing amount of time refining it. The Ops Plan had a long distribution list, and copies went into the team's "Blue Book" that everyone was encouraged to read, though few besides Dari or Juan ever bothered. Mostly, his teammates sniffed at the plans and wondered why John spent so much time pecking at the Mac.

He diverted his anxieties by soliciting proposals from documentary film makers who wanted to make a movie about Daedalus. At times, members of the team didn't know whether a film was being made of the airplane or whether an airplane was being made for a film. In fact, John couldn't always explain why he'd sought proposals from film makers. But he did fantasize about producing a documentary and proposals floated in for more than six months from independent film makers across the country. The producer in Boston already had taped hundreds of hours of the construction and flights of Emily. *National Geographic* had sponsored an eighteen-minute piece on the flight of the Light Eagle. But John wanted more. He wanted to see the flight appear on NOVA, the public television science show. For months, he searched for a sponsor and interviewed producers. He even wrote a script. But NOVA seemed uninterested, and none of the sponsors wanted to fund another Daedalus sideshow.

Team members thought John's fantasies would dwindle. Instead, they discovered that money budgeted for engineering was being shifted into a separate account for a documentary. How could such a diversion be justified, they wondered, particularly at a time when flight testing in California neared collapse? It was unbelievable.

It was a magnificent pantomime. The film script, the ops plan, the many presentations somehow allowed him to anticipate contingencies and plot the team's course. Most of all, the new ventures let him once again hold his ideal up like a lantern and see it reflected in the eyes of others.

So he'd fly to New York for interviews. He'd drive to a series of meetings with Olympic Airlines executives, wearing his black, pin-striped suit and the tie with the tiny Monarch airplanes on it. He'd spend a day with a reporter and photographer from *People* magazine in California. He'd fly back to Washington for a meeting with a documentary producer and *National Geographic* editors, and spend an evening at home. He'd leave the next morning for San Francisco to visit Shaklee, going on to Stanford to give a speech at a conference on high-altitude aircraft. Then to Atlanta for a public relations appearance, and down to Orlando for another appearance at EPCOT. If he grew distant from his team, he was increasingly invigorated in the company of strangers.

After his day in New York, editors at the news magazines alerted a handful of photographers and writers in the Mediterranean to schedule the Daedalus flight on their calendars. A producer at Cable News Network edited John's taped interview. A couple of publishing houses called the literary agent about a Daedalus book. When John left Penn Station, he felt buoyed once again.

On the train to Boston, however, he met a young woman with three children, the youngest of whom looked like Ellis. The children's father had kicked them out of their house in Virginia Beach and now the young mother was moving them to live with an aunt in Rhode Island until she could find a job. The littlest boy told John they were only going to eat peanut butter until their mom found work.

As John chatted with them, he began to think about Barbara and Ellis. He thought about the difference between riding the Yankee Clipper through backwater New England towns and flying the Pan American Shuttle into New York the day before. The airplane had an air phone in every row; the passengers all wore business suits. A couple of brokers with Merrill Lynch sat next to him on the airplane, and when the air phones didn't work, everyone complained. Looking at the three children, he felt a hollowness intruding on his journey. By the time the kids and their mother got off the train, he was depressed.

* * *

By the end of January, John had managed to hire more secretarial help and lured Peggie back to the project. He sent Steve Bussolari and Lois out to California to regain control over the flight operations. NASA agreed to extend Daedalus Project operations on the lakebed through March. He and Ethan planned a long-duration ergometer test for the pilots, using the new endurance drink Ethan had developed. And on Sunday afternoons, he discovered, Mark Drela had started riding his bike out to Hanscom and was working quietly on the leading edge of the second Daedalus airplane. Despite the setbacks of December, the team stayed on schedule.

John also made an effort to fly home to Washington more often. On Saturday mornings he fixed pancakes with Ellis and took his two-year-old son to the "real zoo" to see "real monkeys." He and Barbara took dance lessons at night, and on Sunday mornings the Langford family went to church.

But there could be no real return to a normal life. In early February he revised the Daedalus flight schedule one last time and set a date with the Greek Air Force to provide a C-130 cargo plane to carry two Daedalus airplanes and Emily to Iraklion on February 23. Then he and Gup met with Jack Kerrebrock in Cambridge to discuss delaying the flight until autumn.

"A lot of people are watching this project very nervously," Kerrebrock said.

John laughed. "They'd be even more nervous if they really understood the situation," he said.

The next day, a sunny Saturday, he was back, walking Ellis to a playground near their home. In the quiet of his tree-filled Alexandria neighborhood, with his child's hand in his own, desperation and isolation set in once again. The Daedalus airplane had logged only four minutes of flight time in California. United Technologies had threatened to scrap its funding because they felt he and Peggie weren't providing pilots for public relations appearances and had missed opportunities to have the company's name mentioned in *The Wall Street Journal*. On February 27, in just twenty-one days, Daedalus had to be in Greece.

John took Ellis home. He went into his office and closed the door. He switched off the lights and turned on the stereo—louder and louder. Then he laid down on the floor and buried his head in his arms until his face became warm and wet with tears.

The next morning, Ellis climbed into his parents' bed, carrying a storybook, and shook his dad. John awoke, startled, left breathless by a dream that he had crashed in a jetliner.

22

Erik rubbed a sore cord of muscle in his knee. Before first light, he and Kanellos would have to be dressed for their inaugural flights in Daedalus. During the early trials on Saturday, February 6, the plane's gearbox had broken and a rudder cable slipped. Dark pools of water on the lakebed created convection currents that unsettled flights. But engineering problems and wet weather couldn't compete with tendonitis. As Erik dressed Sunday morning, his mind was on his knee and three weeks of lost training. He needed a good start today; his flight to Greece hung in the balance.

Every pilot worried about rankings. Although each one had made two-hour flights in Emily during January, John continued to set prerequisites. Lois sent weekly status reports to Boston summarizing their cycling mileage, hours on the flight simulator, and trips to the glider port. They were told to prepare for five-hour ergometer tests. But they'd also discovered a passenger list to Athens with only three reservations for athletes. Over the next five days, they suspected, test flights in the new Daedalus would determine who claimed positions on the final team.

Even Kanellos, whose place seemed most assured, believed John was keeping a private rating system and suspected that one or two pilots might be cut. "I know by my own experience that every test counts," Kanellos said. The Greek's athletic superiority gave him top berth, but his flights had been unnerving affairs that sometimes threatened to end in a heap of Mylar and carbon spars. It was said that Kanellos looked like a bull-rider in the cockpit. Even the Greek awoke Sunday morning knowing he needed to prove himself.

Cold air stung their nostrils. The frozen lakebed glimmered. As usual, a team of engineers waited at the hangar. A few had stayed up all night working with soldering irons, oscilloscopes, and computers, hovering over a failed autopilot. They'd even tried to calibrate Finberg's electrostatic sensors by stringing them to lasagna pans and sections of a broom handle. But as the sun rose and a ridge of gold light appeared over the mountains, they cleaned the Daedalus gearbox and prepared for the plane's second day of trial flights.

Kanellos went first. A black pickup truck led him around the lakebed. Bussolari tried to make radio contact without luck, so he shouted instructions from the cab. Lois and Finberg's electronics crew tracked the flight in a van, monitoring sensors strapped to both the pilot and plane. Two slowly trundling vehicles slipped across the mudflats, escorting a colossal twirling bird that floated between them.

In two minutes, Kanellos landed. "Fix the radio," he said.

Shortly, he was in the air again, this time ten feet off the ground. Gusts tipped the Daedalus' nose up. Tom Clancy slipped off the truckbed and ran underneath the plane to listen to the gearbox. Daedalus flew so silently, any imperfection would certainly speak.

The plane nosed up, dropped, the rudder flipped left weakly. Juan craned his neck out the window to watch. Sunlight played along the tail surface, so he couldn't estimate the angles. "One, one, one, one!" Bussolari shouted, counting strokes to see how fast Kanellos flew. As the sun rose, the sky changed from orange to a cold, clear blue.

Greg kept his eye plugged to the video camera. Sponsors' logos shimmered on the tail. Silver, luminous red, translucent white. Sleek. Icy. He held the camera steady for twenty minutes, watching Kanellos bank, negotiate turns, and slip through thermals. His own flight, just the day before, had been flawless. Warm air rose up off dark water spots and bobbled the wings. Greg could see that the Greek had improved.

"Nose up. Ease it down, Kanellos," Bussolari said. After twenty minutes: "Keep pedaling—wing runners! . . . Fine job, Kanellos."

The Greek brushed tangled blond hair out of his eyes and stepped onto hard ground. The engineers trotted out to hear his impressions. Bussolari noticed that he'd flown less aggressively than before, perhaps a sign that he'd gained confidence. He was smiling, too, just the sign Bussolari had wanted—they couldn't afford to disturb delicate foreign relations in Athens with an unhappy Greek cyclist.

While the engineers chatted, Erik climbed in the cockpit and asked Juan to watch the rudder. He twisted the controls. Looked fine.

A slight wind picked up over the lakebed. Although it was getting late, Bussolari decided to let Erik fly. A tour around Rogers Lake might boost his confidence, too.

The takeoff was perfect. Erik experimented with various speeds to find the most comfortable setting and his heart rate rose to 160, 170 in the turns. Just nerves. He flew Daedalus south to drier ground.

"Nice, Erik," Bussolari said. "You're a little left-wing low. Correct with the rudder."

Juan watched the wings flex. He leaned over to cut off the heat in the truck, and cracked a window.

"As you can tell, it's extremely slow coming out of the right bank," Juan said. "We're running out of lakebed here; tell him to give it left rudder."

The turns took much longer than Erik expected. Five minutes to do a 180, and the left wing kept wanting to drag like a lame appendage. He listened for the voice—Bussolari usually spoke in the calmest, most reassuring way—but below he heard chattering. Juan was still asking about the rudder. Watching the deflections, Juan had noticed that the tail didn't flex as dramatically as it should have. He instructed Bussolari to pull the truck under Daedalus so they could watch Erik's turn. The truck eased under the tail boom and everyone glanced up at rear control surfaces.

Glenn listened to radio exchanges from the van. Fight, flight, or faint are the usual reactions in a crisis, he thought. Kanellos was a fighter. Erik tended to hesitate. But today he'd seen the Greek take more pains to let the airplane fly itself, and Erik was making dramatic corrections to pull Daedalus through apparently stiff thermals. As much as he'd come to dislike Erik—was it competitive zeal or arrogance?—Glenn liked to see good flights, and this one looked fine. Maybe as they all gained more confidence the pilots could ease up around the apartment. Competition made life intolerable.

Suddenly, a gust off the lakebed wrenched the airplane 20 degrees off course. Erik's heart rate jumped. The left wing tipped slowly, and the airplane dropped to within three feet of the dirt. Erik locked the rudder control left.

"I think we'll terminate soon," Bussolari said, calmly.

Daedalus rose and dipped, and Bussolari ordered Erik to aim the airplane toward the green NASA hangar, to pull the nose up, to pick up the pace. They had to move past the wet spots. But the airplane kept slipping to the right, and the wing angled down.

"Full left rudder!" Bussolari suddenly cried. "Full left rudder. Full left rudder. Pull on the full left rudder!"

Juan gasped. Greg let the video camera slip.

The airplane quivered, and slowly the right wing tip drooped and scraped across the ground. The cockpit angled up and shuddered. But the plane flexed so well, even as its wing rode across the surface of the lakebed, that for a moment it seemed possible that it would lift up again and fly.

Erik pinned the rudder control and pedaled wildly. The spar above his head cracked, the right wing collapsed. The Mylar windows ripped and popped, carbon tubes and Styrofoam wing ribs snapped and shattered. He burst through the plastic window and leaped out of the plane, diving, arms wide, onto the ground. Erik tumbled across the lakebed, and when he came to rest he curled up on his knees and laid his wet forehead on the dirt. He couldn't look up.

Crew members spilled out of their vehicles. Greg Zack trained the SONY Video Camera on the scene momentarily, then he dropped the camera on the seat and trotted out on the rust-red lakebed, too. Bussolari bent over and touched Erik's shoulder. Erik stood, quickly brushed himself off, and walked straight out into the flats, alone. He could only imagine what they saw: a crumpled fuselage, snapped wings, shredded plastic . . . It felt like slow motion, but he couldn't find an excuse. He couldn't explain what happened. "I just destroyed a million-dollar airplane."

Bussolari examined the wreckage and put his hands on his hips. Juan picked up a shard of carbon and glanced away. The air was sunny and dry and windless.

In Washington, John and Barbara were dressing Ellis to take him to a preschool open house. In Boston, Gup and Mark rambled around the Daedalus airplane, preparing the second, so-called B-ship for final construction. A call from California, about 2:30 p.m. Eastern Standard Time, stopped them cold.

Gup hung onto the phone for quite a while without speaking. He took four pages of notes, and when he hung up, he called a meeting of the team. "This will sound much worse than it actually is," he said.

His notes included an exhaustive inventory of broken parts. Claudia Ranniger looked at the list of damaged carbon tubes and went to work immediately in the solar home. Another UROP, Peter Neirinckx, who thought he'd cut his last piece of pink foam, strapped on an air mask and started again. Steve Darr joined Claudia unwrapping frozen sheets of graphite epoxy. One of the new office workers, Mary Chiochios, arranged to have a copy of the video tape of Erik's crash flown to Boston. Mark rushed back to his office at MIT to recreate an analysis of the airplane's dynamics. Dari and Gup decided that if someone could drive the broken airplane back to Hanscom immediately, they could rebuild Daedalus and still meet the flight schedule for Greece in three weeks. Optimism reigned.

"What we should do," Dari said, "is just have them put Juan in the back of the trailer with a case of Cokes and a ton of M&Ms. Let Clancy drive, and when they get here in a couple of days, we'll open the doors and there it'll be. A new airplane."

Erik called after the meeting, and in a wavering voice asked to speak with Gup. They hadn't seen the wreckage or helped load pieces onto the back of the black truck. Erik knew how bad it was.

"You're a part of this project, Erik," Gup said. "It's an experimental aircraft and you don't have to take the blame for a crash just because you were flying it. Look, I've been in crashes before."

"I feel like I've let the whole team down," Erik said.

"Don't worry about it," Gup said. "We're behind you. You don't have to explain anything to us."

Erik sounded shaken. Gup told him to sort through his feelings and write them down, take a break today, but don't stop training. Get back in Emily as soon as you can. We want you to start flying again.

"I just don't know," Erik said. "It's not like getting back into another bike race after a crash. I don't mind taking a risk for myself. But this is different. So many people have put so much effort into this airplane. I mean, we were about to wrap it up this week and pack up for Greece. And now this. That's a lot of weight. It's an enormous responsibility."

On the flight from Washington to Boston, John reviewed the notes he'd taken that afternoon from phone conversations between Lancaster and Lexington. They're sounding brave, John thought. Juan and Mark think we won't even slip schedule. That's naive. It's the same brave talk I always hear from these guys. Pride. And Gup wouldn't even agree to fly Erik back to Boston because he doesn't want to "disrupt" his training schedule. Wake up! We need to talk to Erik. We just destroyed the airplane, Guppy. Don't you get it? There *is* no training schedule. And he thinks we'll be back in two weeks flying. I'll bet you a week's pay on that one, Gup. I never should have hired Erik Schmidt.

Peggie walked into the hangar that night, back from a day-long flight from Greece. A red and white Santa Claus cap dangled from the rudder of the B-ship and voices echoed in the hangar. She found John drafting a press release and saw messages that United Technologies had called, requesting that the project not make any public announcements about the crash. The company also insisted that any existing photographs of the crash should be carefully edited so the United Technologies logo would not appear.

What crash? Who crashed? she asked. The Associated Press called, then a radio station, then a few newspapers. Peggie was immediately thrust into action.

They worked all night. The next morning John ordered bouquets of roses to lighten the mood around the office. Gup found an old video tape of human-powered airplane crashes—winged monstrosities that humped along pavements and tumbled off cliffs—and played it to amuse the construction team. By the afternoon, when Bussolari's accident report appeared over electronic mail and the video tape arrived in Boston, the Hanscom crew was prepared to face the worst. Twenty-five hours after the crash, the East Coast team gathered in a conference room to watch the disaster. They played the tape over and over. They argued and bickered.

The report suggested that the weather had played only a minor role in the crash. There'd been several cool, dry days in California lately, and all the pilots, including Erik, had logged long, increasingly skilled flights in the Light Eagle—pilot error didn't seem to be the problem. With only a week left on the schedule for flight testing, they'd all wanted to fly the new Daedalus A-ship on the lakebed. Maybe he had rushed it, Bussolari admitted. Maybe it was too windy. In any case, after viewing the tape, he wrote, he felt certain that the airplane didn't break because of gusts or pilot errors or thermals. He suspected the engineers had failed.

The tape showed Erik's entire flight. The right wing scraped the lakebed and the cockpit tilted. When the wing snapped, the tail boom cracked instantly and the fuselage tubes exploded. Everything broke at once—a perfect measure of how well the airplane had been built. Daedalus wasn't too strong anywhere. "Well, at least it's gratifying to see how well it failed," Gup said.

They replayed it. Bussolari's memo pointed out that a cable from the cockpit to the rudder hadn't allowed adequate deflection of the tail surface in turns. Apparently, a change to a lighter grade of Kevlar cable allowed a slight amount of stretch in the fibers, and slackened the controls. There didn't seem to be enough tug in the rudder control to make a safe turn. But even that explanation didn't satisfy the Boston crew.

Bussolari's report didn't mention that the wings had also looked very flat during the Daedalus test flights. But as John, Gup, and Mark reviewed the tape, they saw for the first time that they had made a major miscalculation during construction. The horizontal curl in the wings, from the center sections to the tips, looked even slighter than they'd allowed on the Light Eagle. This curvature—technically known as dihedral—was supposed to prevent the plane from skidding around turns, to let it roll comfortably through the air rather than sideslipping toward the ground. Daedalus suffered instability, they guessed, because someone hadn't allowed for enough dihedral. John grabbed a set of photographs of Daedalus in flight and figured that the height of the dihedral, from the wing's center section to the tips, was about three feet. If they had allowed just a few more inches in the lift wire—the single external wire used to allow the spar to bend—they could have increased the dihedral and stabilized the airplane.

"I guess we mismeasured it," Mark said. "We're off maybe a foot and a half."

There wasn't time to assign blame. The real arguments occurred over solutions. While Gup and Mark insisted that they simply increase the dihedral, John fought for ailerons. He wanted the engineers to design and install new, moving control surfaces (like the flaps on a jet) to the tips of the Daedalus wings. Ailerons, like those on Emily, would provide direct roll control, an essential defense against gusts.

"Of course it will take time to build ailerons!" he argued. "Of course, ailerons will mean more mechanical parts! But you're taking a major risk without them. We need more control on that airplane!"

But Mark and Gup would not yield. Ailerons, they argued, added weight, complexity, and construction time.

It was satisfying to see them fighting over engineering principles rather than personal grievances. Although a gang of UROPs sat around the table listening, there were only three participants in the exchange—Gup, Mark, and John—and they flung technical language at each other like epithets. They seemed invigorated, not defeated. After three hours they were still fighting.

Finally, early that evening, the meetings at Hanscom ended in a compromise. Gup agreed that his construction team would make provisions on the spar to install ailerons. At the same time, the senior engineers would analyze their data in Boston and renew the data-gathering test flights in California. If Gup and Mark felt confident that the airplane could fly with rudder-only controls, they'd have to prove it in the air and on paper.

As Juan made plane reservations to bring back a data analysis to MIT, Gup made arrangements to move out to California and take over flight operations. Nothing could be left to chance.

The next afternoon, John was still sitting in the office working the phones. He had a cold and his eyes watered, but he sounded cheery when he called Barbara in Washington—"We have great slides of the crash," he said.

While Peggie talked to a reporter with *The New York Times*, John asked one of the office clerks to call Lancaster and check on the pilots. Everyone's fine, the clerk reported. Kanellos is in the middle of the first five-hour endurance test at the base hospital. Erik flew today. Everything's great.

"Five-hour test!" John said. "What's this five-hour bullshit! Jesus! Call them back! I've gotta talk to Ethan."

When John finally tracked down Ethan at the Air Force Base hospital at Edwards, Kanellos was well into his fifth hour on the ergometer. John was astonished—he thought he and Ethan had agreed to test every pilot for six hours. Ethan, who'd flown from New Haven to give the tests, balked when he answered John's call. Bussolari, who'd joined Ethan to administer the tests, argued, too, that a six-hour test could stress the pilots beyond their limits; it would increase the probability of injuries, disrupt their training schedules, and provoke dissent. You're changing the rules on them in the middle of the game, they told John. They've been training for five-hour rides.

John insisted.

"We test spars to one and a half times the normal load," he said. "And you guys want to test the engine at a negative margin? I can't believe it! . . . Take it to six. No more discussion. No more debate. You tell them, if they don't try it, they don't fly it."

That night in California the pilots argued bitterly with Ethan and Bussolari. Frank had been plagued with tendonitis, like Erik. Would he risk an injury? Erik, plainly, couldn't take the six-hour test. Did that mean he'd be dropped from the team? Glenn hadn't ever passed a four-hour test. How could he be ready to take a six-hour test in just a few days?

But John wouldn't allow for options. With only two weeks before they were scheduled to leave for Greece, he wanted every part tested and every body scrutinized.

23

A cup of coffee kept Jack Kerrebrock quiet, but not calm. His friends at United Technologies sensed disaster. A few professors in his own department were questioning the value of the Daedalus Project. Somehow he had to regain control.

The Colonial Inn in Concord slowly filled with its regular morning traffic, and Kerrebrock ordered another coffee. He glanced again at the agenda John had sent. Not even at NASA had he encountered deadline dilemmas like this: the crash, pilot training, financial problems, sponsorship conflicts. At seven-thirty—late morning by his estimation—the project's managers appeared. John, Dari, Peggie, and Steve Bussolari wandered in looking surprisingly cheerful.

"We're in good shape for the rebuild," Dari said, as he pulled up a chair next to his former professor. "We have enough chrome Mylar to cover the fuselage. We've made fifteen new ribs for the wing sections and the leading edge is already done."

"Did you check the length of the lift wire?" Kerrebrock asked.

The managers laughed. In the ten days since Erik's crash, they'd done far more than lengthen the lift wire and increase the dihedral. NASA engineers and a young German aeronautics specialist from MIT had reexamined every physical property of the airplanes. Finberg had filled Emily with more miniature data recorders than NASA put on its F-18s.

"Jack, from an engineering point of view," John said, "the crash was the best thing that's happened in a long time. People have been up all night making solutions and pulling out simulations that haven't been used since the Eagle."

"Are you telling me there wasn't an up-to-date training simulation of the

Daedalus airplane?" Kerrebrock asked. He hadn't realized the pilots were riding a flight simulator programmed just for Emily's dynamics. Maybe that explained Erik's crash.

"Don't worry, it's just been updated for Daedalus," John said. "But that's just my point—now we've got far more data analysis to improve the pilots' simulator. A lot of those stability and handling issues have been thoroughly covered."

Kerrebrock's mouth turned down in a distinct frown. He looked at Bussolari. How'd the Light Eagle feel flying without ailerons? he asked.

"All I have is five minutes of experience," Bussolari replied. "But it felt remarkable."

They sounded so optimistic.

"Look," Kerrebrock said, "I need more than this. I want an explanation of what we know and what we don't know. Frankly, I'm a little surprised to learn that the simulation of the Daedalus didn't get done properly. You could assume that the energy pumped into it since the accident will take care of any problems, but I think that's a dangerous assumption."

Bussolari knew how to handle Kerrebrock. "Jack, we've got the data," he said. "What would you like, a report or a meeting of the minds?"

"I just feel like we're assuming someone's on top of things," Kerrebrock said. "But with two airplanes being tested out there, you're going to have a lot of difference of opinion about the data. And I'd like to know when to rely on just one person's judgment. It's not obvious to me how decisions are being made now . . . For example, John, I'd like to know if you're leaving the decisions up to Gup or Mark, without any explanations?"

"It's tough," John said. "You know, we have people on two coasts right now."

Kerrebrock folded his arms.

Bussolari leaped in to offer more support. "I'll be going back to California in a few days," he said. "Maybe we can get on the phone and talk about it then."

"Okay," Kerrebrock said. "Well, that solves one problem, I guess. You know, I'm mostly concerned about the engineering."

"I wish I could agree with you about that," John said. He glanced at the next item on the agenda: pilot training.

The good news, Bussolari said, was that three of the five athletes had passed six-hour endurance tests, and Ethan's lab results showed that their glucose levels remained high throughout. The pilots had complained that Ethan's designer drink was still too salty. But it didn't give them diarrhea anymore and no one suffered from dehydration. Ethanol was a success.

Unfortunately, Glenn hadn't trained sufficiently and failed the test. Erik was still recovering from tendonitis and couldn't risk taking the six-hour test yet. "He's in the most tenuous situation," Bussolari said.

"Well, I'm afraid pilot selection isn't our worst problem," John said suddenly. He slipped a set of graphs out of his briefcase and passed copies to Bussolari and Kerrebrock. The graphs were strung with densely plotted lines, showing estimated flight times versus distances, times versus wind speeds, head winds versus tail winds. Kerrebrock looked startled.

"What's this?" he said.

"Six hours is not the worst case," John said. "Statistically, we're guaranteed a head wind. Eight hours is not unimaginable."

"I don't understand," Kerrebrock said, looking slightly stunned.

"Just look at the data, Jack," John said. "We've got to factor in Murphy's Law."

"Murphy's Law is not statistical," Kerrebrock said, sternly. "Murphy's Law is fatality."

Bussolari scowled and scanned the chart, too. He'd seen these before. He suspected John was using them for political advantage, to bolster his demands to make the six-hour test a prerequisite for the pilots. "If we're facing an eight-hour test over the Aegean, I don't think we should proceed," he said.

"Besides that," Kerrebrock said, "I think loading all this onto the pilots is a big psychological burden right now. To tell them it could be eight or nine hours at this point would be a real problem. The number we've told them all along is five hours and they've trained for five hours and it is still the best estimate."

"That's the dead-air estimate," John said.

"I just don't think you can keep asking the pilots to absorb these problems," Kerrebrock said. "The weather, the time delays . . ."

"Well, the real point is whether we can accept someone on the rotation who hasn't gone beyond six hours in the ergometer test," John said. "I haven't heard anyone yet say that Glenn can do it. And I think we should decide how to look at it. We could say it's like a spar test and if Glenn doesn't make it, he's not going to fly. The other way is to say, Glenn, if you think you can make it, we'll put our faith in you."

"I don't think we can leave it up to Glenn," Bussolari said, "But I also don't see why we should decrease the pilot team down to four."

"It would make a difference to me," John said. "I think we could really use Glenn in other areas of the project—in test flights, public relations. . . ."

Bussolari had grown weary of these arguments. Human beings weren't airplane parts. Each of the pilots had different strengths, but psychologically they were not bearing up to continually increasing demands from the East Coast. He was afraid they could still lose Kanellos to the Tour of Greece race; Frank hadn't flown enough to prove himself a worthy pilot; Erik was ready to quit after the crash. He tried to defend the athletes, but John was relentless.

Kerrebrock listened to them argue, and then he intervened.

"This crash and the six-hour test business have pointed up significant lapses," he said. "And I think we should begin having senior staff meetings, like this one, with some frequency. It's time we tightened up on the hierarchy of this organization. John, I think you have a serious issue here. You should have meetings with these people and then you should make decisions."

"Me decide?" John said.

"Yessir," Kerrebrock said. "There has to be some mechanism for arbitrating and you have to do it."

With that said, Kerrebrock gave his first order of the day. "For example, it's time you decided to bring the sponsors onto the team," he said. "And you can begin by telling everybody to mention these corporations when they're talking to reporters."

The table fell silent. The managers braced themselves.

"Let me make a little speech," Kerrebrock said. "I'm not completely prepared to, but I think it's become necessary. The problem is attitude. There is an attitude on this project that the sponsors are the enemy. Now I want you to see to it that these people become friends of the project. Believe me, this is a serious problem, and I'm hearing about it continually."

Not again, John thought.

"Why don't we just be blunt," John said. "We're talking about United Technologies."

"That's not correct," Kerrebrock said. "I've heard it from Shaklee, and it was there with Anheuser-Busch. The attitude is that this project is some kind of holy circle that cannot be interrupted, and this attitude has to stop!"

Kerrebrock just didn't understand what they'd been through, John thought. Conflicts over logos, publicity, and control had been constant sources of irritation. The latest struggle over who owned the crash photographs—and whether to release them to the news media—had ended in screaming tirades over the phone.

"Jack, it's just natural to have an adversarial relationship with the sponsors," John said. "Each company has its own agenda and a different image to maintain. But we've got a responsibility to keep this from turning into a circus act."

"I'm trying to be constructive," Kerrebrock said hotly.

"These people don't own us," John said.

"That's exactly right," Kerrebrock replied, "but I'm talking about a spirit of cooperation. You can be firm with people and still be friendly with them. I think you need to treat the sponsors the same way you treat each other."

John laughed, cynically, "Not the way we've been screaming at each other around here."

The final topic addressed the project's finances. John and Kerrebrock talked, more cordially, about the deals and negotiations that had broken down during the last few weeks. The budget would finally come up short $100,000. John estimated, however, that spending had only been 80 percent of budget as of February, so if the repairs and tests could be completed soon, the team could be in Greece by mid-March and not miss the calm weather promised by meteorologists.

"Of course," he said, "if we slip and miss the weather window, we'll need some money to go back in September."

"There is no slip, John," Kerrebrock said. Until then no one had gauged the depth of the professor's concern. Kerrebrock was serious. "At some point, you have to face it. At the end of May, there is no more project. You're not going to hold it together."

John looked shocked. His teammates sat perfectly still. "I guess this isn't an appropriate issue to discuss right here," he said.

"But it's always in the back of my mind," Kerrebrock said. "And I want to keep it there."

The meeting ended without resolution. Kerrebrock had forwarded his own agenda, and left them with a threat. But he wasn't finished. There would be more calls that day from the corporations, more mediation sessions. The man who'd been the engineers' mentor for ten years drove back to Cambridge, sensing for the first time that he had a potential disaster on his hands.

Late that night Kerrebrock returned to Lexington, and on the stairs of the Hanscom hangar, he demanded that John and Peggie make amends with United Technologies. If they wouldn't meet United Technologies' demands over the Daedalus photographs, Kerrebrock said, he'd advise his contacts there to absolve funding.

"Fine," John told him, "we'll find another sponsor."

But within days, John and Peggie struggled to renew relationships with their sponsors. UROPs and engineers were encouraged to wear the corporate logos and credit the corporations when talking to the media. John tried to instill more enthusiasm in the team, and issued a memo asking for comments about naming the two Daedalus airplanes "Daedalus '88" and "Daedalus '87," like the America's Cup sailboats "Stars and Stripes." But the engineers ignored him. With the new flurry of orders, they suspected John had sold out to commercial interests. They plastered decals on the two airplanes, officially christening one "A" and the other "B."

In California, the pilots reviewed their contracts with legal counsel, and refused to sign the documents. One of the primary Greek sponsors, Olympic Airways, threatened to withdraw its support because of inadequate press coverage. Public relations officials at United Technologies and Shaklee called Bryan Duff at the Air and Space Museum in Washington and asked him to communicate on their behalf with the Daedalus Project. Whatever relationship once linked the corporations and the management team now ceased to exist.

Just a few weeks before the weather around Crete entered its seasonal calm period, the Daedalus team was divided, splintered into unmistakable factions. Who knows whether any one of them regarded signs of spring—

the lonely purr of a Piper Cub above the hangar, melted snow drizzling through gutter spouts, the Hanscom guards chatting over Boston Red Sox pre-season stories. Only the UROPs remained far enough removed from the disintegrating core to wallow in the remaining enchantments of Daedalus. They worked on the hobbled A-ship at Hanscom night and day, and when John left for California for the last time, they played around the broken airplane as if there would be no flight, no risks, no more troubles. One night, the youngest UROP, whom they nicknamed "Punky," fell asleep beside the airplane and, like Lilliputians, his co-workers lashed him with Kevlar and glued him to the floor. They stopped working to play a video maze game called Shadowgate, and programmed the foam cutter to slice out a wardrobe of pants and shirts made of pink Foamular.

Another night, they broke for an event they called the Epoxy Connoisseur's Nosing Contest. Grant slipped into John's office, grabbed John's blue blazer, and emceed the event: "This contest will decide, once and for all," he said, "which engineer knows his glues." Blindfolded UROPs were led, one at a time, from a conference room to a twenty-foot table in the hangar, where they sniffed fourteen different fragrances and identified the familiar toxins by name: the obscure Southern Sorghum; L-28; Titebond wood glue; Zapagap; DHP quick-set epoxy; Formula 2; Cyanoacrylate Instaset, known as "zip-kicker"; 3M 77; Thixotropic; Structural adhesive; Contact cement; White Microfill; Pink Microfill. Despite spectators' efforts to fool the contestants by waving hot pizza between their noses and the sticky samples, Punky beat Dari by guessing eleven correctly.

They ate at Greek restaurants, sampled ouzo, and ordered fine wines on nights out. When *Les Misérables* played in Boston, half the team fell under its spell. Dari and Grant competed to read Hugo's bulky novel in their spare time (between 3:00 and 9:00 a.m.). Grant also kept a copy of Robert Pirsig's *Zen and the Art of Motorcycle Maintenance* alongside as a companion while he worked—slowly—to build a more exquisite fuselage.

The solar home spilled over with the flotsam and jetsam of communal student life. On a wooden post, where they'd written team members' phone numbers, someone tacked an engineer's drawing of the wing spar, penciled a long calculation over the design, added triangles, arrows, circles, marked the edges of each shape with indecipherable algebraic equations, and at the bottom penned this note: "Any questions, guys?"

A marker board appeared next to the fuselage with a list of "Martyr Points" for those who'd made the greatest sacrifices to the project. Steve Darr gained a couple of points for giving up flight training school to rebuild Daedalus. Punky earned a point for breaking up with his girlfriend, then scored negative points for leaving the hangar one day to visit his parents. Dari, who believed a scoring system spoiled the ritual of sacrifice, refused to accept any points at all, and erased every mark under his name.

They hung a calendar on a far wall, noting the commencement of a death pact. Under a heading "Neglectable Items," they listed: "Fix car, social life, sleep, nutrition, United Technologies logo on airplane, pack for Greece, UROP course preparation."

The undergraduates had a wonderful time. They didn't know that in Cambridge and Lancaster, sponsors wanted out and pilots threatened to strike. The Daedalus Project was moving closer to collapse almost daily.

24

Erik draped a skull-and-crossbones above the kitchen table. At dinner they talked about gunning down Greg during flight ops. They started a rumor that the propeller would fall off during Glenn's flight. At breakfast, Erik told Frank he snuck out at midnight with a garden hose and doused the taxi paths with water: "Too bad, buddy, no ops today." On training rides, Frank pulled up to Erik and edged him over toward the curb. "Oops," Frank would say, with a nudge, "there goes another pilot."

They called them "Dead Pilot Jokes." Sometimes the wisecracks cut too deep, and silence filled the apartment. "It's not like we're all great drinking buddies here," Frank would say. "Face it, this is a giant chess match."

It wasn't just paranoia, either. At their final press conference in Los Angeles the five cyclists showed up wearing logo-infested team jackets. But the press release said only three of them would be in the final rotation in Crete. Who's going to Greece? asked the Associated Press. Must be a typo here in this press release, said the guy from CBS. The pilots turned to each other, speechless. When they asked for an explanation, John was noncommittal. He was not yet convinced they would all make the team. He might take three, he might take five. No, he did not know what criteria would be used. Yes, he did sanction the press release.

"Hey guys," he said, "the Mercury astronauts didn't know which of them was going to make the first space flight until a few days before the launch. If they could handle it, so can you."

One night, with just two weeks remaining before they'd leave for Greece, the pilots received an ominous memo from Boston: the charts John had shown Kerrebrock revealed that a head wind could increase their flight time from five to eight hours.

> As you can see, the nominal five-hour flight time is itself a myth, depending as it does on still air. Further, this assumes the aircraft is flown at a constant fifteen mph speed. If you fly it slower (which the Daedalus A-ship was being flown in flight tests) your times are more sensitive to winds. This translates into a range of distances between fifty and 130 miles! So you can see that it is really almost meaningless to talk in terms of a "seventy-four-mile flight" or quibble about a few millimeters on the map. The actual flight depends on the wind speed and the flight speed. Now if you really want to get depressed ask Lois to show you the wind speed and direction frequency data!

The data looked familiar—Glenn's computations batted back six months later. Worse than the message, though, was the memo's tone. Even from three thousand miles away the Daedalus management exerted a subtle brand of psychological pressure, and as usual, it worked.

After four months in isolation, the athletes' training schedules now approached Olympic levels. Before coming to California in October, Glenn's highest weekly workout on a bicycle had been two hundred miles. After three months, he was doing four hundred thirty miles. Erik and Greg logged hundreds of flights in glider planes. Kanellos and Glenn rode the ergometer constantly.

They also obeyed a strict diet, designed by a nutritionist from Shaklee, Les Wong. During the fall, Les had analyzed each pilot's eating habits and found their diets astonishingly poor. Greg Zack was the only athlete Les had ever encountered who could eat a pound of cheese at one sitting—twice a day. Glenn's after-dinner plates spilled over with cookies and ice cream. Kanellos consumed so much feta cheese, greasy meats, and olive oil that his diet was nearly 50 percent fat. The pilots' diets looked worse than the average American's. Shaklee put them on notice.

"Nutrition will not be a limiting factor," Les told them. "But if they're saying the flight could take six hours, we may be able to prolong your endurance for thirty minutes. That literally may mean the difference between flying and falling. A better diet won't make you Superman, but it will give you that edge."

Ah, the edge. They were soon consuming up to six thousand calories a day. They ate like pigs: 65 to 80 percent carbohydrate (rice, whole grain breads, beans, cereals, potatoes, fruits, and vegetables); 10 to 15 percent protein (lean cuts of turkey, chicken, cottage cheese); and an extraordinarily low 10 percent in fat. They dined between rides, after flight ops, at press conferences, during meetings. Their bodies ran like dynamos.

Not long after they received John's depressing charts, Bussolari dropped by the apartment. His eyes were droopy and dark. His wife was in the hospital in Boston; doctors suspected meningitis. He had to rush through a heavy agenda before he left for the airport: the pilots needed at least a full hour each in the Daedalus B-ship; they needed a new seat in the cockpit, more data from trial flights, and a private shakedown test for John.

"Langford?" Erik asked.

"Yeah," Bussolari said. He held up the latest memo from his electronic mailbox. "John's set a new schedule and wants to come out here next week to watch the B-ship fly. This is really an improbable series of events."

"There's no way to be ready in Greece by April 1," Kanellos said. He leaned over Bussolari's shoulder.

"We still need a lot of work," Bussolari agreed.

"What about the cargo plane?" Glenn asked. "Did John say anything about the delay?"

"Apparently, the Greek Air Force has agreed to change the date until we can get the A-ship rebuilt. But that's it. We've got two weeks."

No one said much after Bussolari left. Glenn climbed on the ergometer and trained for his second six-hour test. Erik dug through plates of lentils and lettuce. Kanellos watched television—the '88 Winter Olympics. A computer printout, almost twelve feet long, looped along a wall like bunting. It wished Glenn a Happy Valentine's Day. Mary Chiochios, one of the office workers at Hanscom, had developed a crush on the pilot before he left the East Coast. None of the rest of them had so much as heard a woman's warm whisper in four months, and the Valentine message, which had hung there for weeks, was a constant, dismal reminder.

At ten o'clock, Lois dropped by with the latest weather report. A cold front was blowing in from the southwest. Flight ops would begin again Tuesday morning at 5:30 a.m. But by Sunday, NASA predicted thirty-knot winds and heavy rain, just in time for John's visit. "You guys want a wake-up?" she asked.

Kanellos and Erik were mesmerized by the women's figure skating on

TV. Whenever a skater swirled and her skirt flipped up over her rear, Kanellos poked Erik. "Go, Erik!" he'd say, and Erik feigned a leap toward the screen. Greg was watching the video monitor on the flight simulator, pulling the cartoon airplane out of slow spirals at the edge of a cartoon mountain. Glenn was in the kitchen now, fixing a late snack.

Wake-up call? Fine. Sure, sure. We'll be there. G'night, Lois.

The final dry days for flying were coming to a close, and the pilots had at least one more order of business to arrange. They'd have a surprise for John when he arrived, whether it rained or not. After Lois left, they stopped their separate pursuits and held a long, private meeting of their own.

Shirtless Frank spread his arms wide and ran in circles around the lakebed. The concentric dance grew tighter as he mimicked the February 29 report in *People* magazine: "The pedal-powered craft hits a sudden updraft . . . banks twenty degrees and corkscrews slowly . . . gracefully . . . irrevocably. . . ." His knees wobbled and his body wriggled like a snake. He spun suddenly. "Full right rudder!" he yelled. "Full right rudder! Full right rudder!" Frank hit the dirt, rolled, and somersaulted backward.

Erik watched from afar. "Only minor damage!" he shouted.

"Wanna bet?" Frank said. He fingered up a flat clog of dirt and offered it to an engineer. Look, he said, we could cover it with caramel and sell it to the company—"Shaklee Lakebed Bars!"

The skull-and-crossbones hung from the back of Emily's fuselage. Gup sat on the gate of a moving station wagon, gripping a nylon cord tied to Emily's nose, dragging Glenn and the airplane like a kite through the air. Emily had lifted up easily and rose to one hundred feet. John's most recent memo specifically said not to fly Emily higher than twenty feet. Gup didn't care what John said.

Besides Steve Finberg and his electronics team, the flight ops crew had expanded by two. A guy from NASA was now collecting data with a young German engineer named Siegfried Zerweckh. Siegfried, who had just enrolled at MIT, had built human-powered airplanes in western Europe as a teenager—he brought an uncommon level of sophistication to the data collection systems. While Glenn flew, "Siggy" monitored a heavy bank of machines in the van: altitude, sideslip, airspeed, rudder control, stability, pitch, roll, yaw—every move Glenn made excited an electronic pulse below. A video camera strung to the tail boom shot into two mirrors carefully angled down the wings, watching the full flex of the new dihedral.

Gup's appearance in California had revived West Coast operations. He gave orders, but he also had fun. The team loosened up under the command of an engineer who showed up for flights wearing a multicolored T-shirt emblazoned with Albert Einstein's image and an inscription that read: SURF REALITY BUT DON'T CUT BACK. He kept rotating pilots in and out of the cockpit, teasing and cajoling them. Kanellos now flew like a jet pilot. Glenn showed incomparable finesse. Greg made one gorgeous flight and waited to make another. Only Frank was still cautious.

"Chase truck ready!" Gup shouted into a walkie-talkie. "Lois McCallinopoulos driving. Greg Zackalopoulos piloting."

Data from the tow tests showed that even with increased dihedral, Emily could bank only at a fifteen-degree angle, an indication that Daedalus might not be safe without ailerons.

But Gup's eye was more perceptive than the data. His stare was a steady, expert analysis of control and stability.

"Awesome, awesome," he said, as Greg maneuvered through the sky. "That guy's a crackerjack . . . He's a cracker, anyway."

Greg bit his lip and strained to keep Emily afloat. The airplane was not only heavier than Daedalus, but Finberg had loaded it with so many electrical devices that the power requirements escalated enormously. The mass of wires and circuit boards above Greg's head looked like a squirrel's nest.

"My legs feel very bad," he radioed.

"They still look good to me, sweetie," Gup said.

"C'mon, mo-fo!" Siggy shouted from the van. "Stop making excuses!" Siegfried had been instructed to address his partners on the project, as well as officials at NASA, as "mo-fo," American ghetto slang whose meaning was never revealed to him. He was wearing surfer shorts and Ray Bans—the newest member of the project was being Americanized under the influence of Gup's hip crew.

"There's a lot of lactic acid in my legs," Greg said. "My rate's one hundred seventy-three."

"Stop making excuses, mo-fo," Gup said.

After a dozen flights, the engineers rolled Emily back to the hangar.

The pilots vanished. Kanellos straddled his bike and disappeared down a road that curled up into the mountains. Greg slipped away without comment. Glenn, Erik, and Frank loaded their gear into the station wagon and headed for home.

The sun was beating down so hot that the pilots drove with the windows cracked and their shirts off. The radio played a loud, thrumming Delta blues. "My confidence is about shot for human-powered flight," Erik said. "I feel like half an engine." Flat miles passed, and no one replied.

A few miles outside Lancaster, Erik pulled the station wagon over at a bar and grill and let Glenn out. Glenn snatched his bike and pedaled off.

When Frank and Erik got home, Frank reached for a beer. He wanted Erik to take a two-day trip with him to Mexico. There was a bike race just over the border, and he thought maybe they could relax a few days, escape from the Daedalus Project before John came to visit.

They argued off and on through the afternoon. Finally Erik said, flatly, no. What if the other racers had seen the *People* magazine article?

Frank exploded.

"I've gotta get out of this place!" he yelled. "I can't stand it anymore!"

He grabbed the car keys and disappeared. It was days before they saw him again.

The engineers met at the NASA cafeteria to discuss the morning's tests. Siggy's analysis suggested that Glenn's elegant turns were really hazardous maneuvers.

What's wrong with the data? Gup asked. He was sure the flights were fine.

Nothing's the matter with the data, Siggy said. Let me have a few more days testing Emily and I can tell you precisely what's wrong with the airplane.

It's no good, Siggy, Gup said. We don't have a few more days. You spent six hours working on it yesterday and I still don't trust those charts.

Two days, Siggy said. I can have all the flights analyzed and it will be perfect.

Well, it's garbage now, Gup said.

Siegfried rose from his seat, slammed a Coca-Cola can down on the table, and walked out.

Tom Clancy, who was drinking a lemonade, absently spit seeds around the patio. Even Finberg, who wanted more data from test flights before putting Daedalus in the air again, agreed. Gup was right. There just wasn't time.

The rains returned Saturday morning. Frank called from San Diego, where he'd met up with a few old friends. John called again to ask why the Daedalus B-ship hadn't flown yet. One of the UROPs lit out on a motorcycle, leaving his girlfriend compulsively cleaning the engineers' apartment and baking cookies to cheer herself up. Siggy was at NASA asking for a $70,000 grant to continue experimental research with Emily. By the time Parky called from San Jose most of the team had vanished.

Parky's involvement with the project had dwindled substantially over the past year. His name was still prominent in news releases, but he rarely visited the East Coast anymore, and even while Daedalus flew in California, he never drove down to watch. Perhaps the project had grown too big and too self-important for his tastes. His priorities had shifted more toward his family, and while engineering continued to fascinate him, the tedium of flight ops and team politics was an annoyance. The project had changed so much over the years—after working on three airplanes, it seemed almost like a business to Parky. And he'd grown tired of the assembly line.

But Parky still had one final, crucial job. Over five months, he and a friend from Lockheed worked every night for up to eight hours in a garage perfecting the Daedalus gearbox. He'd simplified the design on a computer, cutting the weight by 20 percent, reducing the number of screws from more than one hundred to about fifteen. Milling the intricate pieces turned out to be his greatest challenge.

After weeks of working two shifts—at Lockheed and at home—Parky was exhausted. One morning at the office, he stepped out into the hallway, turned pale, and suddenly lost his balance. His friend saw him stumble and reached to hold him up. Parky blacked out. About five minutes later, paramedics arrived and found him stretched out on the floor. A couple of hours later, he was resting in a hospital, worrying about the gearboxes again. He would not admit that he needed more sleep. He told people he'd just tripped over something, that his blood sugar was low.

"I am basically a klutz," he said, when people called to ask what happened. "That is the first principle."

Gup went to visit Parky, to pick up the gearboxes and check on his old friend. When he got back to Lancaster, it was still raining. By five o'clock, rain flooded the streets outside the pilots' apartment, sluicing down Division Street, washing through the spokes of Tom Clancy's motorcycle.

Gup entered the apartment late in the day planning to work a while on the flight simulator. Parky had complained that he hadn't been paid for overhauling Emily's gearboxes. He hadn't been paid for a technical paper he'd written the year before on the flight characteristics of Emily, either. "Don't be surprised if the gearboxes blow up suddenly," he'd said. But the gears looked exquisite, professionally crafted, as clean as the workings of a watch. Gup had driven them back to the hangar for immediate installation into the Daedalus.

As he worked on the Silicon Graphics program, sitting in the fading light of late afternoon, he stopped suddenly. He was tired, too. Parky's face flashed in his mind. He saw lines and crow's feet. Jowls, thinning hair? Years of their lives had vanished. Youth had passed.

"Jesus," Gup said. "Parky's getting old."

The next week John flew to Lancaster and watched Glenn pedal Daedalus for twenty minutes over the lakebed. Film crews wandered on and off the flats scrutinizing the flight. A documentary crew from NOVA, which John had finally managed to lure into the project, made its first visit. Greg Zack took the plane up for a ride and a cable jumped its pulley. He lost rudder control, and quickly landed before the airplane took a dive. The winds picked up, and most of the engineers argued with John not to fly again that day. Kanellos, their least experienced pilot, was waiting to step in.

"If this airplane's going to Greece without ailerons," John said, "I want to see it fly now." Take the worst pilot, put him in the windiest conditions, he said. Prove to me that it can stand up.

John rode with Lois in the lead truck and watched Kanellos cycle, trundle across the ground, then lift up. The airplane hit one gust, then another. The wing fluttered and danced. One tip flipped up, the other dipped toward the ground. It was an unnerving ballet. But the airplane flew beautifully with its high sweeping dihedral. John let out a shout, and clapped Gup on the back. A success!

With just one more order of business to complete, John could return to Boston and put his worries to rest. In a week, they'd leave for Greece.

But that night, in a stormy meeting, the pilots defied him. All five returned the revenue-sharing contracts. They would not agree to make MIT their sole representative after the Daedalus flight. They would not clear any commercial ventures with MIT. They would not refer all requests for interviews or public appearances to MIT. They would not continue to do public relations for the sponsors after the flight.

The previous week, the pilots had rewritten the contract many times, talked to a lawyer, held meetings and debated among themselves about what they could accept. Erik wouldn't sign; Frank wouldn't sign; Glenn wanted to do what was best for the pilot team; Greg was simply sick of the whole affair. But since the pilot selections hadn't yet been made and they were afraid to speak individually to John, they appointed Glenn as their negotiator.

Frank's comments to Glenn were blunt: "If I say I want to do a center-spread in *Playgirl* and they say, no, we want you to do *National Geographic*, who intercedes?"

"*Playgirl?*" John said, when he met with Glenn. "That's exactly the point!"

Glenn would not reveal the positions of his teammates. But the next day, John pulled Erik and Frank and Greg aside. By the time he left California, he was confident that he'd have signed contracts even from the holdouts within the next week. After all, each pilot knew if he didn't sign a contract, he wouldn't go to Greece. That, they were told, was the final prerequisite for joining the team.

John never seemed to recognize the depth of the dissent or, if he did, to consider how it might backfire. Soon, he thought naively, we'll be together on the island of Crete, waiting to fly.

25

The final meeting of the Daedalus team in the United States took place on the evening of March 8 at the Colonial Inn in Concord, Massachusetts. The event coincided with the last day of operations in California, after two hundred thirteen flights of Daedalus A, B, and Emily. John treated his East Coast crew to a fine dinner of fish and roast beef. The UROPs ordered expensive wines, which made several of them drowsy. As the guest speaker, a classicist from Hope College, gave the first modern interpretation of the Daedalus myth ever presented to the team, some youngsters snoozed.

By early March 1988, the only truly sanguine aspect of the project that remained was the engineering. The team had built three excellent human-powered airplanes. During the last days of duress, the engineers painted the airplanes' propeller tips pink. They referred, cynically, to the flight as "The Stunt," and fantasized about it ending in a spectacular crash on the black beaches of Santorini. If Daedalus didn't crash upon landing, Juan guaranteed he'd be on the beach waiting with an ax.

Juan, in particular, wanted to see Daedalus destroyed at the end. Like all the other senior members of the team, he was tired of the project's politics, squabbles with sponsors, and divided loyalties. They'd devoted too much time to this dream to let the adventure become a stunt. In their minds, it was hard to imagine the Daedalus flight being more than a public performance for the sponsors and the media. This dinner, they suspected, was just the beginning of the final spectacle. That night whenever commercial rhetoric intruded on the conversation, Juan let his jaw slacken, his body tremble, and his eyes water. He leaned back, drew an index finger under his nose, and like a kid sneezing a muffled "Bullshit!" during English class, coughed up a boisterous "Shaklee Shake!" The UROPs roared.

John, in his blue blazer, looked out of place among his crew. The undergraduates' clothes held the lingering odors of a dozen glues. Grant tied his hair in a pony tail. The UROPs wore flight jackets and jeans. Most of John's professional peers were absent—some, like Ethan, were too busy to attend, and some, like Parky and Gup, were thousands of miles away, still working on the few remaining puzzles.

Who knows why Juan showed up? The UROPs teased him, affectionately calling him "the Cruz-ter," "Dr. Cruz," "Motley Cruz." He shook his long red hair and peered at them, half-smiling, over his glasses. The project had grown too large, but it had never worn away his loyalty to the mission. It was the myth that kept him there. As the night progressed, it became apparent that his knowledge of Daedalus iconography and Mycenean history reached depths that approached a scholarly understanding. Their speaker, Dr. Jacob Nyenhuis, had come prepared to give a detailed presentation and show a couple of carousels of slides. But a voice in the dark kept interrupting with informed questions that caught him unaware.

"The bird in this painting," Juan said, pointing to a slide of Pieter Brueghel the Elder's famous sixteenth-century scene, *Landscape with the Fall of Icarus*,"—this must be the nephew, Talos. In one version of the myth it says that when Daedalus killed his nephew, his soul turned into a partridge and he watched while Daedalus buried Icarus. That seems to be what this scene suggests."

The professor had never noticed the bird. He looked. Sure enough, out there with the plowman, shepherd, and fisherman who glanced at Daedalus flying overhead, a bird perched, calmly watching the fall.

"Well, I just don't know," Dr. Nyenhuis said. "Perhaps I should look at this again." The professor gazed at the screen.

He showed more paintings, by Rubens, Van Dyke, Picart. When the slide cart finally inched around to the twentieth century, Nyenhuis entered his element. The myth of Daedalus had captured the imagination of artists and writers more during the last one hundred years than at any other time in recorded history. But flight wasn't the pivotal symbol of the age. Rather, it was the labyrinth, the maze. John felt a sudden pique.

The British sculptor, painter, and writer Michael Ayrton had examined the myth through a lifetime of work, the professor continued. This was the subject of Nyenhuis's forthcoming book, and as he spoke of the artist's novel, *The Maze Maker*, and flashed through slides of his sculpture and paintings, the professor began to plumb disturbing metaphysical ideas. Large images of Ayrton's work appeared on the screen. Mazes, minotaurs, labyrinthine skulls. Flaccid, pot-bellied wingmakers. Thick, sagging torsos trying to gain lift with outstretched wings. Flimsy, brittle wings tied to the arms of skeletal human shapes. The artist's life had been an obsessional journey into the myth of Daedalus. At one time Ayrton had identified with Icarus, another time with Daedalus, another time with the Minotaur. John watched, wide-eyed.

"Seeking the archetypal craftsman, Ayrton sought himself," Dr. Nyenhuis said. "Himself a prisoner in the maze as well as its maker, he saw himself in his fellow prisoner, his fellow prisoner in himself."

The room was silent. John was deeply moved. The theme in Ayrton's novel—that Daedalus considered Icarus a "hero" and therefore a fool—played in his mind. Is the contrast between Daedalus and Icarus—between creator and hero—also the difference between engineer and pilot? And the Minotaur, who wanders at the center of the maze, who is he?

After the lecture, Juan returned with the UROPs to Hanscom, where they continued to make a few final repairs on the Daedalus A-ship. But John stayed and talked with Dr. Nyenhuis for a long time. He borrowed a copy of *The Maze Maker*. When he finally got back to the project office, he had to take the phone off the hook. He couldn't concentrate. He went home to bed. The next morning when he drove back to the office, John felt overwhelmed again. He worked for a few hours and then left without telling anyone where he was going. "It's a coping problem," he said.

The next evening, he escaped Hanscom and headed home. He still had memos to prepare, calls to make, potential sponsors to visit, three dozen team members and three airplanes to send to Greece. But on the flight from Boston to Washington, John looked out at the clouds over the bay and pressed his face up against the window. Tears welled up. He was happy. He was empty. He was afraid. When the airplane slipped through the clouds and the glow of the sun reflected off the wings of the DC-9, he basked in a strange warmth. What would he do when this project ended? Where would he go then?

FLIGHT

26

If ever there had been a labyrinth designed to cage the human spirit . . . Athens was as cramped by the excesses of modernity as any maze an engineer could devise. The noisy chaos of an exploding population and an intricate network of narrow streets thick with concrete high-rises crowded the city's jewel, the Acropolis. Even there, where Daedalus reputedly first tendered his skills, carbon monoxide from automotive exhausts threatened to turn one of the hand-hewn wonders of the ancient world to dust. Archaeologists had erected scaffolding around their prized temples, struggling to preserve stone that had turned black and brittle. The gods gave way, it seemed, to the triumph of mortals.

The cradle of Western civilization rested mostly in the imaginations of young Yankees who expected to time-travel into 1600 B.C. lugging Macintoshes, fax machines, calculators, fiber-optic cables, Kevlar, Mylar, and Izers. But antiquity could best be measured by how long it took them to squeeze through the teeming crowds at Athens Airport, process their bags at customs, and hail a taxi from a congested queue. An hour's wait seemed like an eternity there.

When John arrived with Barbara and Ellis on March 19, he entered a modern Byzantium without his American teammates. Half of them still labored at Hanscom with Gup while the rest had already scattered through Greece, checking weather data and making preparations.

As usual, the Daedalus adventure was not what he'd ever expected. Just twelve hours before, he, Barbara, and Ellis had boarded a 747 *Olympic Spirit* in New York, and during their flight over the Atlantic he had reassured her that they had seen the last of those years of interminable separations.

"We're together starting now, right?" Barbara had said.

"Right," John said. "Absolutely, right. Absolutely. From now on."

Barbara had sneezed and coughed; Ellis, who was also sick, whined and struggled in his mother's arms. John looked around the cabin, full of Greeks and Greek-Americans. "I never envisioned it like this," he said. "I mean, all these years I always thought the whole team would be here together whooping it up."

But the separations would continue, at least until the final flight, and his teammates were nowhere to be found. At least the skies were clear and sunny. During the preceding week, the weather had been temperamental, snowy one day, blustery another, placid and warm the next. Bussolari's twenty-year collection of weather data should have prepared them for a turbulent transition from winter to spring. But experiencing the variable climate gave him a scare. It was certainly less predictable than New England, stranger than the desert.

The Langfords met Jamie Pavlou, the team's Athens office liaison, at the airport. She escorted them to the VIP room for coffee and orange juice. There they met John Poutos, the acting director of the government's foreign services division, a fifty-three-year-old bureaucrat in the Ministry of Information. Jamie, a young, dark-featured woman, handsomely dressed, had been the project's primary contact in Greece for months, arranging logistics for the team's arrival. She was cordial and chatty. But Poutos, a stern, stocky man with a heavy brow and thick-framed glasses, seemed nervous. He slung a short, silver string of worry beads in intricate loops with one hand and puffed a Kent with the other.

It was no secret that Poutos didn't like John Langford. After nine official welcomings since 1985, Poutos had come to detect an arrogance that bothered him, and something else—disappointment, condescension, he never could pinpoint it—that kept him at a distance. He especially didn't enjoy taking orders from the Daedalus managers. Theirs had been an uncomfortable alliance.

They made a careening taxi ride from the airport, past hundreds of concrete balconies. The natural landscape was rocky, sparsely green and dry, nothing like the flat, serene California desert or the enchanting, rich woods around Hanscom. And the quality of the sunlight, though so startling that every crack seemed to stand out on every building, was not at all warm and comforting. It stung the skin, cold and bristly.

Jamie tried to make conversation. She pointed out the Olympic Training Center, where the pilots had been living for a week. The Acropolis, Olympic Stadium, the Plaka—all the sights. John listened, absently. He was reviewing his itinerary—a logistical nightmare. Within the next few days he had to find boats, vans, and money. He was not happy with Jamie's advance work.

When the cab pulled up over the curb in front of the lush Hotel Grande Bretagne in Constitution Square, he looked at his watch and sighed. Barbara and Ellis were sick with the flu. He felt clammy and chilled, too. With little more than a week before the Daedalus Project was scheduled to open its flight window at the Hellenic Air Force Base in Crete, he felt the weight of deadlines would have to be borne, as usual, on his back.

He escorted Barbara and Ellis to their room, and then met Jamie in their new office on the first floor of the hotel. The room overlooked a spacious garden courtyard dense with flowering orange trees and bougainvillea. Elegantly decorated with overstuffed chairs, blue and burgundy furnishings, the sunny office quickly succumbed to the Daedalus malady. Stacked in corners, plopped in mid-room, spilling off desks were press materials, cloth logos, photographs, and news clippings. John sat behind his Macintosh computer and sorted through the latest Operations Plan. He had no time to dawdle.

"I'd say the next thing to do is go down to the training center and get those clowns out of there," he told Jamie. "I'd like to see the pilots training in Crete in a couple of days."

A few hours later, John would begin negotiations for the final $100,000 gift to the Daedalus Project. He would cross the marble floor of the Hotel Grande Bretagne and shake hands with the last of a long series of moneyed dealmakers. Raising funds and meeting the demands of this task remained his solitary duty: striking deals, hiding discomfort with a fetching smile, excusing himself from the family to glad-hand among the weenies.

At 7:30 p.m., a tall, slump-shouldered, rubbery faced man stepped through the gilded doors of the hotel, wheezing and scowling. His black hair was wind-whipped and he was breathless from a long walk. John eyed him and followed Poutos across the marble floor, smile in place.

"We have bad news, heh?" said Timos Stavropoulos, an assistant manager of the Hellenic Industrial Development Bank, one of the largest banks in Greece.

Poutos tumbled his string of silver worry beads from one hand to the other. In the past two weeks, he'd negotiated this deal from a temporary government headquarters across from the bank. Poutos had been assured they would sign a contract as soon as John arrived in Athens.

A representative from MIT's news office, Charlie Ball, introduced himself to Stavropoulos and handed him a letter from MIT executives outlining their objections. "I'm sure we can work this out," Ball said. Jamie joined them, too, then a photographer from the *National Geographic* wandered over.

Stavropoulos had made a simple pitch to his board of directors—give the project $100,000 in exchange for a small cloth patch that said, "Hellenic Industrial Development Bank: Invest In Greece," which project members would add to their jackets. But in Boston, where the image of MIT stood above the noisy concerns of a fledgling band of whiz kids, university officials could not see how they could so sully themselves. It was too commercial. Stavropoulos stroked his chin, removed his glasses and tossed them on a table.

"Looks very grim," he said.

"Wait, wait . . ." John stammered. "What about . . . What about 'The Bank That Helps You Invest In Greece' . . . ?"

" 'The Bank Helps You Invest In Greece'?" Ball said.

" 'We Invest In Greece'!" John said. "Or maybe . . ."

Stavropoulos waved them off and bowed his head meditatively. He lifted a felt-tipped pen and a bit of waste paper from a pocket of his rumpled suit. Hurriedly, he sketched a new design. Using his thumb as a centerpoint on the page, he held the pen tip six inches out and spun the paper around, so that he'd drawn a circle, and inked in the bank's name and a new slogan. "We Help You Invest In Greece," he said. Everyone around the table blinked.

"We Help You Invest In Greece," John said. "I like it."

"We Help You Invest In Greece," Ball repeated. "I don't see how anyone could object to that."

"See!" Stavropoulos shouted. "We have made good—how do you say?—progress?"

"Negotiation?" Ball said.

"Right!" he said. Stavropoulos plucked a small bottle of glue from another sagging pocket and produced a miniature Greek flag, about an inch long, and pasted it ceremoniously to the top of his proposed patch. "You Telefax this, and if MIT accepts, we are in business," he told John. "We have found a solution. Let's have a drink now!"

John's jacket front was littered with patches already: a NASA logo, Shaklee Corporation designs, Olympic Airways insignias, United Technologies Corporation artwork, Amoco's trademark, Daedalus icons. Over a million dollars in promises. Stavropoulos didn't notice until the drinks arrived.

"Wait," he said, suddenly. "You have no room for our patch."

John laughed nervously and glanced down at his coat. "No, no," he said, scanning for space. "We'll put it right here . . . right here under . . . NASA."

John awoke the next day dizzy, coughing, unable to stand for long. Barbara and Ellis sniffled and sneezed. The two-year-old's body finally had adjusted to the time change and Ellis had slept fourteen hours. But John had a meeting with the Hellenic Bank's executives at ten o'clock to discuss their deal. If the corporate sponsors in the United States would just accept the bank's logo, he could finally secure the project's budget. He also had interviews scheduled with *Time* and *Newsweek* reporters, as well as meetings with businessmen who might be persuaded to loan vans and boats in Crete.

At eight, he ordered fresh orange juice and a croissant. By 8:30 a.m. he'd slipped on his Monarch tie, picked up his briefcase, and headed downstairs to the office.

Poutos had called to say the project should shut down over Easter week because military support would diminish. Bryan Sullivan left a message saying he'd installed the weather station on Santorini, but a Greek meteorologist seemed confounded by their selection for the flight route—"The wind never stops blowing here," he'd warned. Jamie called to say she'd been awakened sporadically all night by phone calls from a high-ranking administrator at MIT, who told her the Hellenic Bank's logo could be no larger than ten centimeters in diameter. He'd even encouraged her to sweet-talk banking officials into accepting a more modest emblem for their money. Jamie couldn't bear going into negotiations again with Greek men who stroked her arm while they did business.

Steve Bussolari showed up at ten. He looked haggard after an all-night party in Thessaloníki dancing syrto dances and breaking dishes in ritual celebration. Bussolari didn't notice John's dour expression. He had his own bad news: the latest weather data showed they'd already missed a perfect flight day from Iraklion to Santorini.

The good news was Bussolari had established his own weather-forecasting station in an abandoned pilots' barracks at the Hellenic Air Force Base near Iraklion. Soon he would gather daily forecasts from the U.S. Weather Service, the Greek national weather service in Athens, the meteorologist at the Air Force Base, and the Weather Services Corporation in Massachusetts. Within a week, Bussolari could begin plotting his own charts using satellite data from western Europe and the United States as well as regional data from boats in the Mediterranean. Micro-meteorology—forecasting wind conditions accurately along a scant seventy-four-mile path—would be possible with the aid of professionals linked by phone lines, satellites, and personal computers across the hemisphere.

"That's great," John said, "but Bryan Sullivan says the meteorologist at Santorini told him it's never calm there."

"Yeah?" Bussolari said. "The guy here in Athens says the same thing."

"Oh well," John said. "Too bad."

The immediate problem was determining the pilot rotation. Erik wasn't ready, John said, because he hadn't taken the six-hour test. And then there was the matter of whether they'd fly on Easter Sunday, if the day proved promising.

"A four-person rotation is reasonable," Bussolari said.

"So why don't we put Glenn on last, and just let the other three draw straws. The other option is to say Kan is the strongest and let him go first. Just for political reasons. Being first on the rotation isn't the most desirable, but it sounds like it is."

"So it's Kan, Greg, Frank, and Glenn?" Bussolari said.

"Okay," John said. "Now the other thing is Poutos is saying the military's closing down the week before Easter. I don't want to alienate the Greeks, but . . ."

"I have an answer," Bussolari said. "Just don't make a big deal out of it."

The messages kept piling up. The Ministry of Sports called to suggest a pre-flight celebration with nude acrobats vaulting a mechanical bull—a simulation of the bull-dancing rituals that were popular in King Minos' palace on Crete. A telex arrived from Gup in Boston asking to delay the flight window for a few days so his builders could rest, prepare their support boats, and test the airplanes. Furthermore, while loading the B-ship onto the Hellenic C-130 cargo plane, one of the Greek pilots accidentally bumped into a wing and broke several ribs and the leading edge. March 30, Gup said, was just too soon.

John typed out a curt response: Hustle the construction crew to Greece and be prepared to fly in five days. "If you're so concerned about 'sleep' and 'adjusting to the time zone,' may I ask why you didn't plan to get over here any sooner? You really can't have it both ways."

Late in the morning, Poutos called from the lobby.

"The officers of the Hellenic Bank want to discuss MIT's changes for the logo," he said wearily.

When John reached the concierge desk he found Poutos walking in circles, shaking his head, puffing on a cigarette. They trotted over to the bank, and took an elevator to the ninth floor, where they waited to meet the bank's deputy governor, Iacovos Georganas.

"John, what is this I hear about a flight on Easter?" Poutos whispered. "There will be no press support if you do."

John said nothing.

"Oh," Poutos groaned, "the journalists will be at my door and every newspaper in Greece will say, 'The Ministry of Information has done it again—the bird flies and no press.' It will be a flop."

"Don't worry," John said, "we can find room on the command boat for journalists." Of course John did not yet have a command boat.

"Yes, but we both know that means American press," Poutos said. "What about the Greek press corps?"

John shrugged.

Charlie Ball soon joined them. Stavropoulos met them and personally taped samples of the new logo to their jackets. The patch, six inches in diameter, looked like a manhole cover next to the others. Charlie agreed to handle the negotiations so John could leave to join Jamie on a search for the command boat. They still needed a private source.

"Rentals cost four hundred and fifty to eight hundred and fifty dollars a day," John reminded Ball, on his way out the door. "There are a lot of quick ways to bust the coffers."

A tall, svelte man wearing a navy blue suit ambled into the waiting room. Georganas looked stern. Stavropoulos jumped to his feet.

"I hear there's a problem?" Georganas said softly.

Charlie held out a telex that had just arrived from MIT's executive offices outlining criticisms of the bank's new logo. Georganas led them into an executive suite, where he slipped on a pair of reading glasses and glanced at the paper.

"We may have to reconvene the board," he said grimly.

Charlie, Poutos, and Stavropoulos sat stiffly around the table. Georganas gazed out the window.

"So," he said, "we will have to invent something else."

An hour later, Charlie Ball and John Poutos stood out on the sidewalk in front of the bank. Poutos frowned.

"The bank is just trying to do something good for international relations, for the United States and for Greece," he said. "This is a show of good faith. For one hundred thousand dollars, all they want is a simple patch!"

"I've got to get to Crete," Charlie said. "My wife told me this morning, 'Charlie, if you can make it through this, you should be accepted as an honorary member of the diplomatic corps.' I'm not used to this kind of work. If United Technologies doesn't accept this, I just give up."

At four o'clock John sat in a SONY van outside a hotel in downtown Athens, coughing, blinking furiously, pale and disheveled. He'd spent the day looking for a cabin cruiser to use as a command boat. He'd traveled door-to-door through downtown Athens, climbing dark stairways into dismal third-floor walk-ups, and taking mirrored elevators into the swanky corporate offices of the largest cruise ship lines in Greece.

One small businessman had offered his own boat, if John would just trade him his team jacket. Unfortunately, John couldn't afford to give up his sandwich board. Another man, one of the country's wealthiest shipping magnates, said he couldn't loan a boat during the holidays but he could find other sources. The man called his secretary and excused himself for a few minutes while John and Jamie drank fresh-squeezed orange juice. The businessman made a telephone call from a private line and a few minutes later he returned.

"Don't worry," he said. The businessman had made contacts in the government who guaranteed that a few members of the Coast Guard would be called back to duty during Easter. It was astonishing to see how readily the government responded. John still didn't have his command boat, but suddenly he did have the Coast Guard at his disposal.

A few minutes after four o'clock, John wandered back to the hotel, his blue Oxford cloth shirt hanging outside the back of his pants. He had to arrange one last deal. In the lobby, he met a representative of the Aero Club of Greece, who had agreed to provide official observers for a record-breaking flight. The man escorted John to a bar, ordered ouzo, olives, peanuts.

"I used to build rockets when I was a boy," the man said, waxing nostalgic. He ordered another round of drinks and snacks. Unfortunately, he said, he wasn't prepared to have his members on call for the next two months to verify a record flight. "I'd like to help . . ." he said. John thought he'd settled this deal with the club months ago. They had to have verification to establish a record.

"Don't worry," the man said. "We will have our members in Santorini when the plane lands to clap our hands. If you need anything—anything—call me first."

"Great," John said. "Great. . . ."

With the last major deals apparently made, the project manager enjoyed a sense of momentary relief. They'd have money in their pockets again soon. He called a restaurant in Plaka and made reservations for his family.

Late that night, though, he returned from dinner and found an urgent message. Kerrebrock had called to say the deal with the bank was off. Now United Technologies had rejected the patch. John exploded, raced to the office and phoned Boston.

"Jack, this is so incredibly disgusting!" he shouted over the phone. "Look, I'm over here to run the goddamn flight. I'm not over here to raise money for you. It's so stupid! UTC is a $17 billion operation and MIT is a huge billion dollar operation and they're arguing over a stupid patch? . . . It's just amazing how petty this is . . . I want to know who made this decision . . ."

Kerrebrock threatened to hang up if John didn't calm down. He said United Technologies was trying to compromise by making a new version of the bank's patch. All he had to do was negotiate again over a simpler, less overtly commercial design.

"They just don't want someone else's name on their jacket, is that it?" John said. "I want out of the loop, Jack."

Fine, Kerrebrock told him. He'd call Charlie Ball to finish the job.

When he hung up, John went into a rage. Although he had the title of project manager, his status at MIT wasn't much higher than a janitor. On MIT's payroll, he was officially listed as a visiting engineer. He'd come to be thought of—quite ambivalently—as the boss among his team, but he answered to a half dozen corporate public relations offices and long line of university administrators. He wanted to tell them all to shove it; he had an airplane to fly.

"MIT looks at us as a group of mavericks," he shouted as he paced the office. "We are a group of mavericks! Look at Parks—it took him seventeen years to get an undergraduate degree. And then the deal with Juan and the complaints that this project was ruining his education. And Gup, Gup's a brilliant engineer with a high school diploma . . . And Mark Drela—Drela's a professor who drives through stop signs at 67 miles an hour. Clancy was kicked out of school last year, and I said, 'Great! now we can hire you full time.' These aren't the kind of people they want to stick out in front of the world and say we're the pride of MIT."

But they were being tolerated by the Institute. And the longer John railed, the more he came back to acknowledge his dependence on it. There was no escape. He might have spent three years holding together a group of the most creative, rebellious, and innovative aeronautical engineers of his generation to build a remarkable airplane of plastic and graphite and Styrofoam, but here, he would have to acquiesce. MIT expected favorable publicity. Daedalus, he realized again, did not belong to him or his engineers. It belonged to MIT, and MIT was beholden to its benefactors. This was the way of the world.

The next morning, he would take another telex to the Greek bank and ask again for their patience.

The government of Greece had taken an interest in the Daedalus Project primarily to increase tourism. And when John met with Greek officials—whether publicly or privately—entertainment and publicity values were the currencies of exchange. Scientific, interdisciplinary, cross-cultural references aside, the Daedalus team was there, basically, to drum up business for a lagging tourist trade.

Since 1985, after the infamous TWA hijacking, American tourist traffic in Greece had dwindled substantially. President Reagan had issued stern warnings at a press conference that year telling Americans to avoid Athens Airport; Reagan blamed the government for poor security measures. Political relations between the two nations had been damaged as a result, and a deluge of airline cancellations from the United States cost the country's number one industry dearly over the next three years.

The Greek ambassador in Washington had originally thought the Daedalus Project could begin to revitalize the tourist trade. At the same time, in the Ministry of Sports, officials were looking to the project to help them with plans for the 1996 Olympic Games. It was said that those in charge supported the Daedalus Project as a way to demonstrate the country's ability to serve as a guest nation for an international athletic event.

After two years of correspondence between the Americans and Greeks, and meetings with administrators in the offices of tourism, sports, and press information, the Greeks were expecting to profit from the flight. But it had to be a media event to succeed. They had offered John the use of their Air Force, Coast Guard, Navy, airlines, meteorological services, and government aides. They would feed and house the project's three dozen team members at a hotel in Crete for up to two months. They'd direct press coverage. They would loan their best cyclist. That was the deal.

But over the past six months, relations between Greek officials and the Daedalus Project had become fraught with tension. The reasons were often petty and personal. John could never tell where Poutos's allegiances lay, and he did not believe Poutos spoke with the authority he claimed to have.

Poutos, he thought, might guarantee the nation's support at its Air Force Base in Crete and the service of its military personnel, but John didn't think it prudent to rely simply on Poutos's word.

At the same time, he began to feel less confident about Jamie's ability to run the Athens office. She was smart, certainly, a former student at the Kennedy School at Harvard, and for six months she had built up a separate foundation of support from the private sector in Greece, just in case Poutos failed them. Of course, when Poutos discovered that the project was striking deals with the private sector, he was angry at the threat to his credibility. Privately, John sometimes blamed Jamie for fouling the relationship.

The situation was further complicated because Jamie was having trouble getting paid. A clerical mix-up between MIT and the Athens office had kept her off the Daedalus payroll for months. After a series of letters through the fall and winter, her financial conflicts with the project still hadn't been resolved and Jamie was angry and suspicious of John.

The pilots, who had been in Athens for a week, fared no better. They complained of jet lag and asked to delay the flight for at least three weeks, to let their bodies recover. They'd also been pestering Jamie to take them to discos and introduce them to some of her friends. After four months in the desert, they felt a serious need for female companionship. Kanellos had been busy with his own girlfriend and didn't have time to show them Athens nightlife. Glenn had met a few women at the training center. But Frank and Erik kept venturing out alone, discovering without exception that Greek women were wary of Americans on the prowl. At discos they couldn't even find dance partners. "I guess we just look too hungry," Erik said.

They were a rangy lot. They'd never resolved their pent-up hostilities from the contract dispute. Greg referred to their crew as "the Daedalus Problem" by the time he left California. Erik and Frank thought Greg and Glenn had broken ranks during the contract negotiations, and left them looking perilously like renegades. That wouldn't help their chances on the rotation. Glenn, on the other hand, had arrived in Greece claiming that Erik and Frank abandoned him at the apartment the morning they left California for New York, which caused him to miss his flight. Erik claimed that Glenn had gained weight and couldn't keep pace with the other cyclists. He couldn't understand why Glenn was assured a spot on the rotation and his own place was so precarious.

"Glenn's extremely different from Frank and me," Erik said, once they arrived in Greece. "He's not an elite athlete, he's a team player. They like him. They don't necessarily like us. We're not the perfect choir boys. And Langford wants choir boys."

At least they all biked together. They toured the countryside, through olive groves and into the mountains. They arranged to ride with Polish defectors and Olympic cyclists from across Europe. At times, when the American pilots felt disgusted with each other and their predicament, all it took was a hard ride with an Olympian to give them perspective. Kanellos and his friends outdistanced them on the straight-aways and left them heaving and swaying on long ascents. Four would-be Daedalus heroes watched jut-jawed Europeans glide up hills without standing.

One morning, Kanellos loaded the lot of them into his black sports car and drove them to his parents' home near Patras. Kanellos sped along the coastal highway at one hundred miles an hour. When they arrived at the Kanellopoulos home, a modest whitewashed house in the country, the Americans anticipated a weekend of relief. They were surrounded by fragrant pine, cypress, and myrtle. The vineyards, green foothills, and the blue Corinthian Gulf promised relaxation. The family had prepared fresh vegetables from the garden.

But even in that isolated village, far from Athens, they couldn't escape the pecking order. The Americans were foreigners now. "Why do you need these other guys?" Kanellos's father said to his son over dinner. The walls of the little house were covered with cycling medals. Kanellos's father laughed and laughed. "Why do you need these other guys?"

Turmoil in the Athens office peaked with John Langford's arrival. Jamie didn't trust John, John didn't trust Poutos, Poutos didn't trust John. The pilots' ranks were ragged. For a week, battles went on, by phone and by telex, between MIT, the Greek government, Jamie, John, Poutos, and the Hellenic Bank. In the cradle of Western civilization, at the birthplace of myth, the Daedalus Project fractured at the moment of flight.

Near the end of March, John's father and mother arrived in Athens in time for Greek Independence Day. They stood on the balcony of the Grande Bretagne and watched soldiers and tanks flood the streets. The Judge listened to John talk about the project's troubles. He was careful not to give instruction or advice.

The elder Langfords would follow their son to Crete, but they couldn't stay. The Judge and Mrs. Langford decided to spend a few days in Santorini and then go back home to Atlanta. They'd have to catch the flight on television.

But the Judge did plenty of thinking. He didn't understand all the political machinations, the discord with the pilots and struggles between sponsors. But he didn't have to know everything. With measured words, he penned this note the night before he left:

> We are very proud of you and we are pulling for and praying for you. Two predictions, however. 1. The most beautiful day will likely be Easter, when Greece is celebrating something else. 2. Mutiny. It comes in every expedition; lasts about two days. Keep calm, don't worry about someone's frank comments or challenges. Stick to the routine. We love you dearly.

27

After a two-day flight in the cold, rumbling stomach of a C-130 cargo plane, a crew of UROPs arrived at the air base outside Iraklion, Crete, with three complete human-powered airplanes. In a cool, gray dusk the Daedalus team assembled under the plane's belly, and waited for the cargo doors to drop so they could stretch their pink wings over Greek soil. Rocky cliffs spilled dramatically from the edge of the runway, sloping into the cold, blue Aegean. A milling crowd of Greek soldiers closed in around the plane. The discordant chatter of broken English and stumbling Greek rang out in the dark.

"It will be impossible!" said Lieutenant Tasos Pavlopoulos, the cargo plane's co-pilot. "I have flown fifteen hundred hours in southern Greece. Winds come over these mountains and . . . intense turbulence. Do you think the ancient Greeks imagined this fantasy? This is like a miracle if a human can do this."

Engineers huddled inside the cargo plane, taking a quick inventory of the freight. UROPs unloaded water skis. One of them wore a sweatshirt from the University of Iceland and another wore a Hard Rock Café T-shirt from Reykjavik, souvenirs from their layover in the north.

Before they left the States, however, there'd almost been trouble. In New Jersey, just before takeoff, the C-130's Greek pilot had carelessly leaned on the B-ship's delicate leading edge and broken a few ribs. Steve Darr, one of the UROPs who'd worked endlessly perfecting Daedalus during its final stages, exploded. He'd grabbed the Greek by his wrists, pulled him within a whisker's distance of his own face, and yelled at him. "You broke our airplane!" Steve cried. "You broke our airplane!" Only in Goose Bay, over repeated toasts to Daedalus, did they make amends.

The media bus, a large, sleek, air-conditioned vehicle, arrived carrying reporters from ten or twelve major Greek newspapers, and the NOVA camera crew from Boston—three guys carrying lights and cameras and sound machines, wearing Red Sox baseball caps—slipped in and out, covering the event like the war on the Western Front. They lit up the airplane with enormous floodlamps. John wandered around under the monstrous gray-green thing with its pregnant belly imagining it was about to give birth on the Greek airfield.

"Wow," he said, "this is like science fiction."

Inside, hanging stem to stern, were parts of three airplanes—wings, spars, props—and office supplies, boxes of food for the pilots, the simulator, computers, apples, T-shirts, press kits. He directed them to carry the parts and baggage to an empty concrete fighter plane hangar just off the runway. When they unloaded the pilots' boxes of clothes and food from Lancaster, someone discovered Frank Scioscia's box full of Pop Tarts, peanut butter, corn chips, Cheerios, and strawberry fruit chewies. "We have to save him from himself," said one of the engineers. At midnight they were still nibbling through Frank's stash. The Greeks and Americans applauded and hugged one another when the last of the Daedalus planes came out the bay doors. The celebration went on for hours.

Barbara and Ellis disappeared with the crowd of journalists and soldiers. They were already in bed back at the hotel in Iraklion before John even realized they'd left. When he returned to the Hotel Xenia with the engineers and climbed in bed, for the first time since leaving Athens a few days before he heard the sea outside. Water washed violently up against the stone breakers, sending a cold spray toward the hotel. Gusts rattled their windows.

The Hotel Xenia sat on a slight promontory, its glassy face overlooking the Aegean. To the west, springtime snows powdered the tips of Mount Kouloukonas. Clouds of dust and salt water layered the horizon, distorting the shape of the island of Dia, a few miles northeast. Even without a forecast, one could look from the glass doors of the Xenia dining room and anticipate the day's weather by watching Dia. On good days, the sun raised a flat, lustrous sheen from the water; the one taverna on Dia stood out as a white speck on a fist-shaped mass. On windy days, an orange mist—dust from the Sahara—stretched a wide band over the water, tinting Dia's face. During days of transition, the island vanished in a gray, swirling fog.

The morning of the day Gup, Juan, and the rest of the team were expected, less than forty-eight hours before the flight window opened on March 30, Glenn and John and Lois held a planning session in the Xenia's dining room. When the meeting began, they had a clear view of Dia surrounded by a tranquil sea. An hour and a half later, the winds outside picked up to fifteen knots and the sea's surface was flecked with whitecaps. Dia shrank from sight.

As the hotel windows began to shudder, the three of them walked over to the glass doors and looked out on a bank of fog and swirling clouds. How could they predict such weather? Clear morning skies might suddenly turn dark.

"Mild weather could kill us," Glenn said.

"I'm absolutely terrified about it," John said.

They were relieved to learn that Bussolari had predicted the unsettled weather. And so it would be for the next two days. For Wednesday the 30th, however, he had forecasted calm. The same for Thursday. Kanellos, it seemed, might be lucky on his first day of rotation and make the flight to Santorini.

At the Air Force Base, about four miles outside the city, Tom Clancy worked against heavy gusts off the water to lash down a corner of an opalescent plastic sheet that covered a temporary hangar he'd built for the Daedalus B-ship. People crouched and slipped through the unzippered front door, careful not to allow dirt to blow inside. Photographers, in particular, wanted to see the new Daedalus lair because filmy white plastic filtered the daylight into a gorgeous warm glow. Building the hangar reminded Tom of the times he'd helped stage some of his father's larger pieces of sculpture. Inside the tent, the air was still and luminous. The long pink wings of Daedalus lay in sections on sawhorses, covered with a dusting of red clay. An American flag and a Greek flag hung, side by side, from a set of aluminum supports.

Gup and the construction crew arrived from Boston just after dinner. John immediately called his weary team into the dining room to announce the schedule. A ferry would bring the last two rubber inflatable boats in the morning and Tom Clancy would be in charge of boats for a drill on the 30th. That left them one day to put the airplanes together, test the boat engines, and prepare a simulation run.

Unfortunately, gas stations on the island had just called a strike, and it wasn't clear where they'd find fuel for their engines. All the pilots would make test flights for the news media at the Air Force Base Wednesday afternoon. And Kanellos should be prepared to fly on Thursday, the 31st. That didn't leave time for the engineers to dally.

Kanellos's head rose up out of the crowd. Then a fist.

"John!" he called. "The rotation has not changed! I am to fly on Wednesday!"

John didn't respond. Tom leaned over to Juan and whispered, "The conflict starts."

"Wednesday is the day of my flight," Kanellos said. "Now you push it one day ahead? If you say I am to be ready the thirtieth, I am ready. Now, you are changing the day to Thursday, but the rotation stays the same. I want to know why. Is the problem of the flight the weather and only the weather?"

"No," John said, at last. "It's not the weather, Kanellos. The engineers need a solid day's work to prepare the airplane. But it's just as if there has been bad weather or a mechanical problem."

"No!" Kanellos said. "The problem is the rotation. I am supposed to fly Wednesday or Thursday. Now I will lose a day."

"I'm sorry," John said. "But the policy is not to change the rotation any time we have a delay. That will play havoc with every pilot's training."

Kanellos pursed his lips and looked down at the floor. The pilots had always argued that once they set a rotation schedule, it mustn't be changed. Kanellos would lose his first day to the engineers.

"I will eat it," he grumbled.

The streets of Iraklion looked like the presentiment of Juan's nightmare ad-venture. Thousands of brass baubles, cheap ceramic pots and vases pressed into the shapes of dozens of Greek icons—bulls, griffins, god-desses, dolphins—spilled from noisy venues, and the smell of greasy souvlaki laced with acrid whiffs of automobile exhaust floated through the streets. The remains of ancient buildings squatted on some street corners, crumbling untended since the bombing runs of World War II, while the modern, low-lying concrete cityscape sprawled from the harbor into an arid countryside, chockablock, colorless. And yet it was a lively town, whose people honored traditions and revered their past. Daedalus was welcomed with a spirit of celebration and goodwill.

Mary Chiochios taught most members of the team to say "Good morning"—*Kalimera*—and "To your health!"—*Yasas!* They sewed the newest cloth patches from the Hellenic Industrial Development Bank onto their United Technologies jackets and sauntered over to Town Hall, dodging motorcycles, peering into fresh vegetable markets, and glancing at Greek fishermen who spilled their morning's catch of octopus into wooden tubs along the dock. Laughing strains of bouzouki music sang from the tavernas.

The team entered the Town Hall and took seats around a large, U-shaped table beneath crystal chandeliers. John gave the mayor a team T-shirt and addressed the assembly. "On behalf of United Technologies and all the institutional investors who support the project in both Greece and the United States, let me say what an honor it is to be here in Iraklion, the birthplace of aviation," he began. His teammates gazed at the dark, ornate Venetian architecture as he spoke. Secretaries from the mayor's office passed out a pictorial history book of Crete and hard-backed Greek calendars, offering the gifts like a sacrament, saying to each one, "We wish you a successful flight . . . We wish you a successful flight . . ."

Two dozen of them left the ceremony for lunch at the Marina Restaurant, on the harbor. A bouzouki band rocked the dining room with pounding drums and blaring trumpets. A team of Greek dancers leaped and tumbled across the floor, slinging sabers. Eight cold bottles of Armanti sat on each table. The commander of the Air Force Base introduced John to the captains of the two Navy torpedo boats that would escort Daedalus on the day of the flight.

Bussolari toasted Ethan Nadel over in a corner with a glass of wine.

"It was, what, Ethan, three years ago that we first met? Your office . . ."

"You were so young, Steve!" Ethan shouted to be heard over the music.

"And I said, 'We want to do this flight in Greece, what do you think?' "

"And I said, 'Well, no way.' "

"Yeah, but then I said, 'Let's at least take a trip to Greece and see. The beer's sixty cents a bottle. . . .' "

"Yeah?"

"And that's when you signed up."

They toasted one another, and then moved quickly through the crowd to toast their pilots, then the engineers.

Although the room overflowed with Daedalus personnel, the veteran engineers—Dari, Juan, Gup, Mark, and Grant—had slipped off to the Air Force Base. Through the quiet of the afternoon, they tended the airplane and talked.

Grant was angry. He'd arrived in Crete only to discover that John's Operations Plan listed him for a tour of duty in Santorini. He would spend his first few days preparing weather reports on the remote island to the north. If the airplane flew soon, he wouldn't see it. After two years building wings and parts, Grant felt betrayed. He was devastated. And he wasn't the only one.

"John's driven by efficiency and productivity," Grant complained. "He doesn't see the emotional side that people have. He doesn't realize that people have been dedicated to this project because they're emotionally involved with the airplane. And if he doesn't account for it, it's got to lead to problems. He can't just keep plugging us into all these organizational charts."

When everyone returned to the hotel before dinner, Gup and Juan tried to talk to John. They needed one more day to prepare the plane; and they wanted him to reconsider Grant's assignment. John said, quietly but firmly, no.

"There's a pecking order here!" Gup said. He pounded his fist against a table and cursed. Had John forgotten the ideals of their project—teamwork, the aesthetic nature of engineering, freedom over bureaucracy? His voice rose, and guests in the bar turned from watching TV to the argument outside the press room. John quickly suggested they take it up again after dinner.

After the meal, the team assembled in the dining room. Gup explained that he needed another day to prepare the plane. Thursday should be a test flight only. Someone had to check the gearboxes; the tow system hadn't been installed; he'd found a push rod in the rudder unlatched that afternoon; he needed a table and a cabinet to store supplies in the temporary hangar; there were so many pieces still sitting in unopened boxes.

John wouldn't yield.

"It's the same problem with any group of engineers," he said. "The closer you get to the day of the flight, the more hesitant you become. You can't let go. This project has always been driven by craftsmanship . . ."

"Bullshit!" Gup said. "That's bullshit." He pointed a finger and narrowed his eyes. "It's driven by your schedules!"

As if on cue, Grant raised his hand and said he wanted to discuss the rotation of engineers to Santorini. Suddenly, the dining room filled with quivering voices. Other UROPs raised their hands to talk, too, and were shouted down by the project's managers, who didn't want to hear any more about sacrifices and Martyr Points and slacking. But the UROPs had spent two years building Daedalus and they wanted to see it fly. One of the young women on the team started to sob. Waiters and cooks peered out of the kitchen.

"I've spent literally months building this airplane," Grant said, "and it's difficult to part with it now without seeing it fly. Don't you understand?"

"No one has a 'right' to see it fly," John said. "And I can't believe you would knowingly do something at this point that could jeopardize the flight."

Gup volunteered to take Grant's place. Juan volunteered. After an hour and a half, the meeting ended without resolution. John could not, publicly, back down. What if he relented every time someone screamed at him? They were too close to a flight. Why hadn't they responded to his Operations Plan before now? He remembered his father's note: Stick to the routine.

The team left the dining hall at ten-thirty, shaken, still arguing. The documentary film maker from NOVA made plans to set up his cameras every night after supper, in case of a mutiny.

Grant sat in the press room to calm down. "I'm ready to sit back and see what happens," he said. "At least people know how much I care about this airplane." Maybe in six months, Juan said, John will realize what a pain in the ass he's been.

John, on the other hand, didn't complain or sulk. He simply recalled an adage he'd heard for years, at MIT, at the Skunkworks, at Lockheed in Atlanta: At some point you have to shoot all the engineers and just fly the airplane. "That's an old slogan," he said, "but it's true."

The team met again at sunrise on the docks. Inside the stone barriers of the harbor fishermen sewed their nets and talked. Their old wooden trollers bobbed at the end of ragged lines in the water. Daedalus crews rushed around three sleek red and black inflatable boats whose engines idled serenely. Their orange rubber Amoco wet suits glistened in the changing light while they waited for a radio transmission that would send them racing out to sea. Commanders on two torpedo boats let their engines rumble, and John, Ethan, Lois, and the camera crew for NOVA boarded the command boat, a cabin cruiser on loan from a local travel agent, Yannis Vlitakis. Shortly after seven o'clock the command came from Bussolari at the air base. It was time for tests. The flotilla seemed perfectly groomed as it curled out of the harbor and in a sudden roar quickly sped toward Dia.

The rubber boats leaped waves like dolphins and the command ship, directed by a computerized LORAN (Long Range Radio Navigation) directional system, set a straight test course fifteen miles into the Aegean.

No one had expected seasickness, but nearly everyone on the command boat vomited over the side on the trip back to the dock. No one on the inflatables wore life jackets, and one of the UROPs was tossed overboard. Steve Finberg's radio transmission system on Santorini failed. Half the hand-held radios on the inflatables didn't work, and most of the team didn't even know how to use them. Nuts and bolts on the boat engines kept loosening during the ride. A couple of the rubber boats almost ran out of gas.

That afternoon, Gup returned to the hangar with a list of problems to resolve. They couldn't rush into the flight window in two days. "I'm not being chickenshit, like John seems to think," he said. "We didn't work for three years on this project to come to this point and fail."

Gup's girlfriend, who'd left her job in California for a few weeks to come to Crete, met him at the hangar and tried to persuade him to join her rock climbing. He needed to forget about John and the project for a few hours.

But it wasn't possible. The boat crews were filtering into the hangar, ready to join Gup and work through the afternoon and into the night.

The next morning the team met the press at the runway. Kanellos would make "the first human-powered flight in Greece since 1600 B.C."—as it had been billed. A gang of photographers and journalists crowded around the Daedalus airplane. Poutos, who was already hoarse after spending the previous day on the telephone with reporters from around the world, gave a raspy introduction: "Today's test flight," he said, "will be historic."

The five pilots stretched and nervously shed their warm-up clothes. Glenn and Greg chatted with Lois. Frank had psyched himself up so much the day before that he rode further than he'd ever ridden in his life. Cycling west toward Rethimnon, he'd pedaled over gravelly goat trails in the mountains. Finally he found his way back to the hotel long after dark. He'd done one hundred forty-five miles, which left him worn and unable to sleep. Erik bowed his head at the end of the runway, facing the sea, and drew himself inward. He would fly today; perhaps he would soon enter the rotation, too. Kanellos's hands and shoulders trembled as he pulled off his sweats. His fingers shook so much that he could barely grasp his shoestrings. Greek journalists called his name. "Bravo, Kanellos!"

Daedalus gleamed. A dozen engineers reached up to grab its carbon boom, fiddle with wires, clean its wings. The blue Aegean sparkled. Glenn plucked a few fine, dried sprigs of grass off the ground and let them drop from his fingers. John watched them fall. Dead calm.

"Windless," Glenn said.

"Scary as hell," John said.

The press gathered around the plane as Kanellos stepped in for the first flight. Even though he'd only make a 200-foot test hop, photographers angled from all sides.

"Get your goddamn camera away from our plane!" shouted one of the engineers. "Move back! move back!" A scuffle broke out. Epithets flew.

But Kanellos hardly seemed to notice. As he entered the crowd tempers subsided. He climbed into the cockpit, and everyone stepped aside. The engineers took their positions and Kanellos began to pedal.

Within seconds, he was in the air. Daedalus skimmed aboveground not more than ten feet off the pavement. Kanellos smiled at the cameramen and turned his thumb up for them to see. The cameras clicked and whirred. John hugged Poutos. Even if Kanellos didn't make the actual flight, at least he'd make the newspapers and television news broadcasts for one day. A Greek had again flown under his own power in Crete.

"When's your day on the rotation, Kanellos?" a journalist called, as the Greek stepped back into his sweatpants and let Greg climb into the cockpit. "Do you start tomorrow?"

"Today is my day," he said, cheerfully.

Later at breakfast all five pilots sat together in the Xenia dining room, looking grim. Les Wong, the Shaklee nutritionist, had given the hotel's cook orders about how to feed the pilots—lots of carbohydrates, some protein, little fat. This wasn't the banquet Kanellos had expected. He gazed at the bowl of oatmeal, the tray of bread and oranges.

"This is the worst a Greek can eat!" he said, dipping his fork in the sticky goo of oatmeal. He called a fat, scowling waiter. "Bring me olive oil, oregano, yogurt! We will mix!" Looking through the glass doors toward Dia, the sea appeared perfectly still. Kanellos could have been halfway to Santorini.

Greg left the breakfast table to call his parents in North Carolina and tell them to take the next flight to Greece. Bussolari had forecast good weather for one of his two rotation days—a 60 percent chance to fly on Saturday, April 2. Erik and Frank spent the rest of the morning talking with Ethan—they argued that Erik's tendonitis had completely cleared up and it was time to move him into the rotation. They knew John would oppose it.

That afternoon, John introduced Greg Zack to an assembly of journalists in the press room. Greg was so anxious beforehand that someone ordered a glass of warm milk to settle his nerves. While cameras clicked and journalists asked questions, Greg sipped his drink. He faced international correspondents from *Time*, *Newsweek*, *Life*, *Sports Illustrated*, Reuters, the Associated Press, *The Boston Globe*, and a half-dozen Greek newspapers. His lips curled up into a shy grin and he swiped off a white mustache. "Flying with unbroken concentration for five or six hours will be difficult," he said. "Powering the airplane and flying it at the same time will be difficult. In fact, I can't see anything that will be easy." Greg bit his lower lip. Greg grinned.

Later, Bussolari set up a marker board in front of the room and drew the first day's weather map. His black isobars and arrows showed the movement of a low pressure system heading eastward from France across the Mediterranean toward Greece. "A southerly flow will bring us rough winds tomorrow," he said, "but by Saturday or Sunday there will be a good possibility for favorable conditions."

After dinner that night, John scolded certain team members for fighting with the Greek press at the Air Force Base. But Brian Duff said he had the problem under control. It wouldn't happen again.

Steve Finberg reported that his electronic communications relay in Santorini still wasn't ready. "I don't know what happened," he said. "It was working in Cambridge."

"Whenever he says that," Siegfried whispered, "little bombs go off inside me." The electronics crew had heard that line too often.

Then Juan suddenly stood up and shouted. "We had perfect weather today!" he said. "These weather forecasts are bullshit! Why didn't we fly? The airplane is one hundred point zero percent ready."

A few people in the dining hall snickered. One hundred point zero percent?

But everyone was too tired to argue. Besides, Gup had a surprise announcement: "I'd just like to say we all had a great flight op today and we should now consider ourselves officially"—a bottle of champagne appeared in his hands, and he lifted it up, gripped the cork with his thumbs, and gave it a nudge; the bottle popped and cold wine burst out—"ready!" The UROPs gathered around Gup to fill their glasses.

Juan, Mark, and Dari retired to the hotel lobby. They'd spent the afternoon at the hangar, fiddling with the airplane. They looked bored and tired.

"So," Dari said. "One hundred point zero percent."

There was a long pause. Mark and Juan stared blankly into space. Dari leaned forward and looked at Juan, then Mark. He slowly combed his fingers through his curly hair and knitted his brow. Then he lurched forward and pointed his finger at Juan, ready to speak. Juan called out first.

"Landing gear bolt!" Juan shouted.

"You . . . !" Dari said, and slapped the table.

Precisely. An MIT engineer does not use the expression "one hundred point zero percent." And if they had sat there for another hour, Dari or

Mark or Juan would have thought of something else they could do to prepare the airplane. For an engineer there is always something to be done, always an improvement to be made, particularly in the final hours when the artist must turn the creation loose. If nothing else, they could always change a steel bolt to a lighter, titanium one.

And in the morning, perhaps they would.

Bussolari's weather forecast for Friday, April 1, came true. Winds at the Air Force Base gusted to forty knots and tore grommets from the plastic canopy on the temporary hangar. Gup and half his building crew worked all morning making emergency repairs, trying to hold the hangar in place. Winds snapped the plastic viciously against the aluminum supports and threatened to lift the entire shelter and toss it into the sea. The pink wings, stretched out along sawhorses, trembled when gusts blasted the tent and stirred the warm air inside.

At the Xenia, journalists cornered Mark and Juan and quizzed them about Daedalus technology. The reporter from Reuters was unusually interested in its long-range applications—particularly military uses, specifically, reconnaissance. He finally put it this way: "Spy satellites?"

"Unfortunately," Juan said, "about ninety percent of aerospace research ends up in military uses. It's something I've been concerned about, personally, for a long time. But right now we don't have anything to do with the military, so it's great." But the reporter wasn't satisfied, and the interview dragged on for almost an hour while the windows rattled.

Jack Kerrebrock arrived from Boston during the afternoon. He checked the press room to make sure the United Technologies logos were hanging on the walls. They weren't. Just an oversight, someone said. Kerrebrock had left MIT without the complete blessing of his colleagues, and he'd grown weary of defending a project that had lapsed so often lately into backbiting and bickering. It was at his behest that Brian Duff joined the team in Crete, because Jack feared the project was headed for disaster. Kerrebrock wanted to see Daedalus fly, and he wanted to be done with it.

Bussolari drew maps again at the afternoon meeting. The movement of a low pressure system in central France was expected to dig south. But tomorrow, he predicted, would be calm enough to assemble the airplane on the runway and do another boat drill. On the other hand, if the weather pattern developed faster overnight, there might be a chance to fly to Santorini.

At 3:00 a.m., doors up and down the third floor at the Xenia swung open, and in the darkness Daedalus crews stumbled out to the elevator and stairwells. Lois awoke the pilots. John was already dressed. It wasn't a drill.

Frank grabbed Greg, who had wandered off into the Xenia dining room, and sat him down.

"Wake up, Greg!" Frank said. "This one's for real."

Greg yawned and massaged his calves. "What's this?" he said. "Saturday? We won't make the morning TV news shows."

John caught a ride with Kerrebrock to the docks.

"I told everybody as soon as I got here we'd fly," Kerrebrock said.

The boat crews scurried around for gas tanks to load onto their inflatables. Yannis Vlitakis, the local tour agent, arrived at his yacht, the *Dulcy II*, wearing a Mickey Mouse sweatshirt, expecting only a short morning drill in the command boat. Greek soldiers on the torpedo boats stood, inexplicably, in uniform, holding rifles. Fishing boats nodded at their moorings in the moonlight.

At five o'clock, Dari veered his inflatable out of the harbor and sped to Dia to take a wind reading. He radioed back: three knots. A few minutes later the Daedalus assembly crew radioed the command boat that they'd begun hitching the wings and prop on the plane at the runway.

The moon had turned pale by the time the order came to head for sea.

"There are too many people on my boat," Yannis told John.

"Don't worry, we'll take care of that," John said.

"But we will take care of it now," Yannis said. "I cannot have eleven people on my boat."

"It's okay, Yannis," John said. "This is just a trial run."

"Okay, but never again!"

Of course, it wasn't a trial run. At 6:45 a.m. the command boat, two torpedo boats, and three inflatables sat bobbing in the water at the edge of the runway. Gup radioed the command boat that he'd gotten no weather reports yet from the team on Santorini. Juan radioed that the wind had picked up over the runway and he wanted to stall final assembly until they could determine if they'd just been hit with a temporary morning land breeze.

"I don't like this," Bussolari said, twisting around through a hatch in the roof of the *Dulcy II*. "The water's murky. We're really being tossed."

Greg climbed into the cockpit, and at seven o'clock the assembly crew taped the Mylar door shut. Juan reported five knots of wind at the end of the runway, just within the margins. The boat crews stood in their quivering inflatables and waited for a signal from Bussolari. They could see the airplane—a tiny dragonfly on the lip of the runway. Without radio transmissions from Santorini Bussolari could only watch the waters around him, follow the course of a weather balloon as it lifted off from the end of the promontory, and talk to the meteorologist from the Air Force Base. Then he had to decide.

"I really don't trust these seas," he said. The water turned silver-gray and waves rolled in ten-inch swells. He pressed a button on his transmitter. "I think we're too late," he said. "Let's scrub it."

"What's going on here!" Juan shouted.

Greg Zack sat in the bar at the Xenia late that afternoon. His parents had arrived just in time for his flight. His sister had flown in from Switzerland. From the Navy boat, where they'd joined the press to watch, they thought Greg had taken off. As they sped through the water behind a series of leaping inflatables, no one bothered to tell them that it had turned into just another boat drill, that the Daedalus hadn't left the runway.

As far as Greg could tell, there'd be no attempted flights during Easter week. Counting forward through the rotation, he could see that his chances of flying had just fallen considerably. Frank would get his turn, then Glenn, then Kanellos, and if Erik entered the rotation, Greg would be bucked into dead time—the holiday vacation. He rested his head on his hands. "This was my last day," he said.

Juan buried his worries that afternoon at Knossos. He explored the ruins he'd longed to see, the vestiges of King Minos' 3,500-year-old palace. Surrounded by olive groves and vineyards, the enormous rocky foundations and crumbling corridors showed evidence of remarkable engineering, stunning architectural skills. Fresh water pipes and a clever sanitation system coursed the building. Juan traced their paths and walked down angling corridors where tiny rooms suddenly appeared, filled with large jugs for wine and grains. He entered servants' quarters, disappeared into dank corners illuminated only by light shafts, discovered a room used for religious rituals, and finally wound up in a gigantic courtyard once used for bull-dancing events and ceremonial sacrifices. At last Juan could actually see the drawings of double-headed axes on the walls, symbols he'd read about for years.

He carried a tour book on his exploration of the palace, but he found he could never follow the map beyond the second story. He kept circling back to the same spot, the queen's bedroom. Lost in the labyrinth, Juan wandered the palace for more than five hours. Soon, the place became familiar to him. So quiet, and before long, dark. Juan caught a late bus back to the Xenia, imagining the flight of Daedalus from the summit of the king's palace. The visit made him feel better. Strangely, it seemed to make him calm.

He went to bed by one o'clock, and two hours later the whole team was awakened again. The morning was quiet and still, and Frank Scioscia was imagining flight.

28

At night Frank practiced visualization techniques from a stack of cassette tapes. ("In just a few hours, this concentrated lesson will allow you to etch, like a powerful laser beam, desirable sensory-rich success-oriented images of achievement into your brain and throughout your entire nervous system.") So at 4:00 a.m., after Lois knocked on his door, he emerged not just as Frank, but as a sort of Super Frank, hauling four cereal boxes under his arms—Raisin Bran, Corn Flakes, All Bran, Rice Krispies. He shuffled into the Xenia dining room for breakfast: four liters of milk, a half gallon of orange juice, a quarter loaf of bread, oatmeal, and his cereals. Someone shouted, "All American Chow Down!" and people peeked into the dining room to watch him eat.

He laughed nervously, told jokes, made wisecracks. "It's a good thing this isn't Easter Sunday. It'd be a shame, a good agnostic like me." He entertained bystanders with a few boisterous burps, and then he disappeared.

Upstairs, John had just discovered that Marc Shafer—the youngest UROP, known as "Punky"—was not in Santorini on weather duty, as assigned, but instead somewhere in the Xenia hotel. Thinking that Punky had deserted his post to see Frank's flight, John stormed through the third floor looking for the UROP. He tore the curtains off the closet in one room and bolted down the stairway calling for Jack Kerrebrock.

"I'm going to rip his head off!" John said, as he paced through the lobby. "Jack! Jack, you have to come with me. You have an important job to do. You're going to save the life of an undergraduate."

As Frank went back upstairs to finish preparing, John met Punky behind the closed doors of the Xenia dining room and held a brief inquisition.

Punky, as it turned out, was innocent; he'd found a replacement to serve duty. "Well," John said abashedly, "given the circumstances you did the right thing, but don't ever do it again."

At 6:30 a.m., Erik joined Frank on the chilly runway, sitting on the ground next to the cockpit. Deceptive land breezes blew down from the mountains to their south. Frank lowered his head and listened to conflicting conversations over the headset. We're gonna fly; wait. Fly now; maybe not. Looks good; don't like this wind. See what happens. They waited for weather reports from Santorini, and the boat crews silently stood in their rubber crafts, listening to the water lap against the sides. They watched the sun rise, red and promising. On the Navy boat, Kerrebrock stood by, waiting helplessly, too. He whispered: "Fly it! Go on! Fly it!"

Bussolari poked his head up through the top of the *Dulcy II* and watched the weather balloon rise fast and straight without wavering. Winds aloft were negligible. But from the command boat, he felt the sea stir. Cool, wet air brushed through his mustache. "Goddamn these winds," he said. "I have no idea where these winds are coming from. We better wait."

So they waited. Erik stayed next to his friend. At seven-thirty, Bussolari again analyzed the data coming from Santorini and the Air Force Base. He asked them to wait even longer. He looked through the brassy sparkle of sunlight on the sea and tried to interpret the significance of underwater disturbances as the water changed from a clear blue to a murky taupe.

"This sea looks strange," he said, finally. "It's a scrub."

The next morning Frank was up again at four-thirty. But this time his eyes were bloodshot from lack of sleep, his hair tousled from a night spent tossing and turning. At breakfast, teammates showed up to watch the show and encourage him to eat Wheaties. "If you make the flight, it'll guarantee your picture on the box," one of the engineers said. But Frank wasn't joking today. He ate a bowl of Grape-Nuts, and then he was gone.

At 6:38 a.m., Frank appeared on the runway again, standing beside the airplane. Bussolari's head again poked up through the top of Yannis's boat. A cool wind blew over the cliffs. An eight-knot wind was reported at the edge of the runway. At 7:07 Frank climbed into the cockpit, and Dari raced his boat out to Dia for a last-minute reading. At 7:14, gusts of up to twenty knots were reported at Santorini, four knots from Dia. Small swells appeared on the horizon.

"I'm worried about this," Bussolari said. "Everybody's giving me bad news."

Navy boats and inflatables looped around the coastline and waited. At 7:26, Bussolari took one final call from the meteorologist at the Air Force Base—winds were rising to ten knots. He lifted his microphone and paused. "Sorry," he radioed the crew, "it's going to be one of those days." And then to John: "Boy, those Navy guys are going to be pissed at me."

But it wasn't the Navy commander who was angry when Bussolari scratched a flight for the third day in a row. Juan confronted him as soon as they returned to the hotel.

"This looks just like yesterday," Juan said, pointing toward Dia from the press room. "It's glass out there. Look at it, it's dead air!"

"I know it's very tempting, Juan," Bussolari said, "but as the sun gets higher in the sky today, you'll see the winds picking up. We'll get a stronger circulation."

"I hope you'll keep track of it today," Juan said. "There seems to be a pattern where we get three to five knots at the takeoff site every day and it gets worse until dawn. But after that . . ."

"I held off as long as I could," Bussolari said.

"People at the airport told me they don't see these weather balloons go up this fast. Ever. Straight up. Dead air."

But it did no good. Even Juan knew the decision might come down to a coin toss anyway, despite all the data Bussolari collected around the globe. Five minutes to takeoff, they had to trust their senses, not science.

The string of delays played havoc with the project. With time to wait, the UROPs went water-skiing. Inter-project issues, like the Santorini assignments, had to be hashed out again. The commander of the Air Force Base decided he couldn't let the team use the concrete hangar any longer to house its Daedalus A-ship. He wanted to close the base entirely the next week so his soldiers could visit their families during Easter. Yannis Vlitakis, who owned the command boat, refused to let team managers wake him up another morning at four o'clock just to be told three hours later that they'd changed their minds. Glenn came down with a sore throat. Frank started spending time with Greg's sister. Curiosity seekers broke into the temporary hangar at night.

Greek journalists also began to ask more questions after the weather briefings—outside the Xenia. How much money had the Daedalus Project cost the Greek government? They observed vacationing UROPs, watched their own weather forecasts, and wondered how clear skies could seem like unacceptably stormy conditions to the Americans. What was the real story?

But nothing could have been more disconcerting than the sudden announcement that the project would no longer pay for the pilots' lunches.

The topic came up one night after dinner. Peggie and John invited Ethan into the press room and made their intentions plain. Unlike any other team members, whose breakfast and dinners were paid from the project's budget, the pilots feasted three times a day, as John put it, "from the trough." John and Peggie said they had heard teammates complaining about favoritism for the pilot team.

Unquestionably, the pilots ate enormous amounts of food, specially prepared meals—stacks of steaks, gallons of fresh orange juice—that weighed heavily on the project's tab at the hotel. Peggie spoke sternly to Ethan about the apparent inequity. John talked, philosophically, about the issue of "entitlements" on the project.

"Most projects like Daedalus are run on the basis where everybody pays their own way," John said. "They're a rich man's game. And a lot of people, like some of these pilots, wouldn't even get a chance to play. But we've run this with an egalitarian philosophy all along, to make it possible for the best qualified people to go, not just the wealthiest. I have to draw the line somewhere. And I'm drawing it here. We have never paid for their meals, and I don't know why we should start now . . . What a perfect symbol, a free lunch. These guys want to take every goody they can get."

Ethan looked aghast. He hadn't heard anyone complain about the pilots' lunches before. How could they possibly consider cutting them off now?

"I'd argue that the pilots aren't like just any other team members," Ethan said. "At this point, I really think you should be looking at them like finely tuned engines. They need six thousand calories a day to function. I don't think you'd scrimp on any other part of the airplane, particularly one that's so vital."

Ethan argued for a half hour. But it was beyond his control. It wasn't his project. The management team called several meetings with the pilots to discuss the issue, but at least one or two of them always refused to show. At last, John made the decision. But the pilots kept instructing Xenia waiters to charge their meals to his tab.

Journalists and photographers at the hotel gossiped about the increasing hostilities that divided the team. The pilots didn't make any secret of their latest conflict with management. The news was inconceivable to reporters. They cut off their lunches? How could that be? Rumors of every color began to fly.

One night, after yet another unpromising weather forecast, John Poutos announced that the Greek press had attacked. A prominent newspaper in Athens had published a lengthy editorial criticizing the project. He called a meeting of the team and cautioned people not to overreact. He displayed the bold headline, "Postponing the Flight of Daedalus" and translated the editorial into English.

The article said the Americans had transformed the myth of Daedalus into a farce, pardoning themselves for not flying by making subtle references to psychological pressures and surrendering to a daily clash of barometric highs and lows. The Greeks were appalled that the project leaders had resisted calling for a flight during the week of their own Catholic Easter, but promised no such deference for the upcoming Greek Orthodox Easter. Commanders and troops aboard the Navy boats, angry about the prospect of missing their families during Easter week—the most important holiday in Greece—were losing patience.

In Iraklion, the editorial said, public opinion had also turned against Daedalus. Townspeople predicted that once the project had satisfied obligations to its commercial sponsors by promoting products in European markets, Kanellos would be dropped in favor of an American pilot. Glenn Tremml, the editorial said, would surely be chosen to fly Daedalus, not Kanellos. Indeed, the rumor was that he had already made private contractual agreements with an undisclosed multinational corporation to be a public relations spokesman after the flight.

" 'The ultimate goal of the project,' " Poutos read, " 'is to serve its interests and nothing more.' "

The room filled with quiet hissing as Poutos folded the newspaper and set it aside. He waved his hands and silenced them. This is serious, he said. We must settle this now.

Most of the editorial had been ill-informed. But they could no longer avoid the issue of whether to fly on April 10, Easter Sunday, the most important holiday in Greece.

"Maybe we could turn it into a joyous occasion," said one of the UROPs.

"How you gonna do that?" another responded. "Mythology is pagan territory."

"You can't fly on Greek Easter," said Mary Chiochios, who'd lived near Sparta as a child. She knew.

"Certainly it's not our choice," Kerrebrock said, "but if Greek Easter is one day out of a hundred to fly, we cannot *not* do it. We have to sit here and wait for Bussolari to make the call. I think the people of Greece will understand that. Do you agree with that, Mr. Poutos?"

"We will do the impossible to help you on that day," Poutos said.

"Maybe we should prepare a way to make it appear as a celebration, one way or another, just in case it happens," Ethan said.

They argued back and forth. There was a pause.

"Have we done anything to have a priest bless the plane and pilot before they fly?" Jamie Pavlou asked. A few people laughed.

"This is not a religious venture," Peggie said.

"I will volunteer to bless the airplane," Dari said.

Gup liked the notion, though. Not only was there something practical about such a gesture but it struck him as refreshing, a chance for the team to learn about the Greeks' passion for their holiest holiday. Poutos supported the idea, too.

"Priests here bless ships before they sail," he said. "Why not an airplane?"

"You can't equate Greek Easter with American Easter," Peggie argued. "I think we should ask ourselves, whether if we were at home, we would fly on Christmas."

A loud chorus of affirmative shouts rang out from the team. Of course.

John realized he would have to visit the archbishop in Iraklion as soon as possible. For more than a year he'd asked his team for an opinion about flying on Easter, and they'd never responded. But if a priest's blessing would help, he'd take his hat in hand and ask. They couldn't afford offending the Greeks now. They couldn't commit sacrilege.

That night Poutos arranged a visit between John and the archbishop. They would meet on Good Friday.

Late Friday morning, as Bussolari optimistically mapped out the progress of a high pressure bubble moving up from Libya to the west-central Mediterranean, a fight broke out in the lobby of the Xenia. A tall Greek man confronted John about several thousand dollars in back wages that allegedly hadn't been paid for the operation of the Athens office. As journalists and team members watched astonished, John shouted at the man and then angrily grabbed him by the arm and told him to "step outside."

Out on a stone barrier, on the edge of the sea, the two argued violently. One of the UROPs took orders to stand by the door, but a gang of people watched the dramatic scene through the glass walls from the lobby. The man shook his fist at Langford. John shouted back. As the argument intensified, people watched to see if one of the men would toss the other over the barrier into the Aegean.

It never turned into a brawl. But when John waded through his team and left the hotel to meet with the archbishop, he knew the scene would create a whole new set of rumors and attacks from within. A serious financial dispute with the Greeks could very easily dissuade the government from continuing its support.

John clutched the steering wheel on the way to the church and ground his teeth. He leaned forward and his shoulders stiffened. He could hardly speak. In his absence, rumors about financial misconduct and soured relations quickly spread through the hotel.

Hundreds of people lined the aisles of St. Minas Cathedral carrying icons and flowers for Good Friday services. John, escorted by Poutos, stopped immediately when he entered the wide doorway of the church. He gaped at the outpouring of emotion. People kissed golden tablets and tossed flowers into a wooden bier. Banks of votive candles flickered in their cups. Women sobbed and reached out to touch the holy shrine. Both men and women wore dark, formal mourning clothes, stiff suits and black caps. John bowed his head and crossed himself.

Archbishop Timotheos met them in his second-floor study, in a building next to the cathedral. Dressed in a somber black robe, masked by a wispy gray beard that flowed to his stomach, the old priest smiled broadly and his cheeks lifted, plump and rosy. John dropped to one knee and kissed his hand. The archbishop seemed delightfully surprised.

"The archbishop wonders if you are Catholic," Poutos said.

"No, Episcopalian," John said.

The priest offered them glazed fruit, orange juice, coffee. John stammered, "Water . . . just water. Water would be fine."

They sat in a stark office on the second floor. It was cool there, and dark. They chatted for a while through their translator about religion, America, Atlanta, MIT. The archbishop said he'd read many articles about the Daedalus Project.

"In my country," John said at last, "good weather on Easter could be taken as a sign from God to fly the airplane. But I am afraid that here in your country, if we flew on a holy day, it might be considered a sacrilege. We also wonder about how the flight could be perceived, given the mythological nature of the project—whether it would be seen as a pagan . . ." he searched for the right word.

The archbishop wrinkled his brow empathetically and waved his hand for John to stop. He looked toward Poutos to translate. Poutos spoke for a moment and Timotheos's face rose in a toothy grin.

"The archbishop says he has written a book on mythology," Poutos said. "It's called *The Seven Glances of Phaestus*. It seems he is quite interested in your airplane and the myth of Daedalus."

John laughed nervously.

"Well, what about Sunday?" John said. "Ask him about Sunday."

Poutos explained the project's dilemma. John couldn't tell whether Poutos raised the subject of blessing the airplane. Again Timotheos smiled and spoke for several minutes directly at John.

"He says from what he reads in the newspapers the airplane is good to revive the myth and expand the frontiers of technology," Poutos reported. "He says, 'It is good to have faith in God, but it also is good to have faith in life.' "

When they left the church, John looked at Poutos, astonished at the reception they'd received. "Did you hear what he said?" Poutos said, laughing. "Someone, write it down!" John replied. The day was sunny; Easter, promising.

But John returned to the Xenia only to find a swelling list of problems. The scene outside the lobby that morning had generated alarming rumors about whether the project had kept its bargains with Jamie and Greek representatives in the Athens office. A web of miscommunications had lingered for six months, and Kerrebrock had been called in to negotiate.

That afternoon, the Langford family moved out of the Xenia and checked into a hotel down the street. Tensions within the team had become too great. Too many people were angry at John. Rumors, which could no longer be contained, grew vicious.

Thick fog and light rains clung to the coastline of Crete through the Easter weekend. Gup spent several hours Sunday morning with a NOVA camera crew, explaining piece by piece how the airplane worked. Early Sunday afternoon, they were interrupted by a gentle scratching on the hangar flap. The air base commander walked in, followed by a few members of his staff, and then a large old man with a gray beard entered, eyes wide, laughing aloud.

"It's the archbishop," Gup said.

He waved off the NOVA team and showed the old man around the hangar. The archbishop, wearing gold chains and a flowing black robe, peered into the cockpit and touched the wings. He talked with the Greek soldiers. The technology amazed him.

"We were hoping you would bless the airplane," Gup said.

One of the soldiers translated, but Timotheos only laughed. He kept touching parts of the airplane. It was so light, so large.

"Archbishop," Gup said again, "as long as you're here, would you like to bless the airplane?"

This time, the old man didn't need a translation. He spoke to one of the soldiers, and the soldier turned to Gup.

"The archbishop says, 'You could pray for good weather, but I cannot bring forces to bear.' "

Not exactly what Gup had in mind. But given the circumstances, it wasn't bad advice.

29

Whether it was sunny or not, their days tended to be enormously complex. More than patience to wait out weeks while the weather grew increasingly warmer and less amenable to human flight, they needed tolerance. The few journalists and camera crews that pledged to stay for the duration melded into the team, and most days, after the weather forecasts, they escaped the hotel together. They slipped away in small groups on motorcycles and on foot, by bus, boat, and bicycle. UROPs asked their weatherman for a daily sun-screen index. By noon, the lobby and press room of the Xenia emptied out. The hotel's clerks and waiters sighed with relief.

John saw his team dissolving, but any effort to bring them back in line was met with scorn. Passions that could no longer be contained were awkwardly cloaked. Gossip and innuendo flourished about sexual liaisons and politicking within the team. Late at night, violent arguments broke out behind closed doors. People could often be seen crying in the lobby or at the bar.

The weather left them in limbo. Bussolari's scribbles in black and red ink over the marker board looked as jumbled as the joke diagrams the engineers once drew to explain the aerodynamics of Mark's airfoil. Arrows and isobars collided with high and low pressure systems carving through Africa, diving from Russia, plunging out of France. As the days grew longer, Bussolari delivered his weather news with an increasingly dramatic flourish. "The good news is the upper air flow from the north and east will bring cooler temperatures," he'd say. "The bad news is it will bring radiational cooling and invite a heavy fog layer tomorrow. In two more days, once these clouds move out of here, we'll be looking for some calm weather."

Glenn called them "the X plus two forecasts"—every day, Bussolari predicted favorable weather in just two more days. It was enough to give the pilots adrenalin poisoning. Juan suggested that Bussolari offer the team a morning seminar, entitled "Meteorology: The Science of Hindsight." One of his forecasts even made the team's sniping "Quote of the Week" list, along with Finberg's famous explanation, "Well, it worked when we were in Cambridge," and John's comment to Punky after his apparent desertion at Santorini, "Given the circumstances you did the right thing, but don't ever do it again." To these they added, "In two more days, once these clouds move out of here, we'll be looking for some calm weather."

Rather than endure these taunts, the weatherman invited team members to visit his makeshift forecasting station at the air base. In a small, chilly, deserted barracks just a few hundred yards from the airstrip, Bussolari worked around the clock drawing weather charts and making phone calls across the Aegean.

He was up many nights until midnight. He napped on a hard bunk, awoke at three o'clock in the morning, and walked outside to test the winds, then went back to his office to phone sources in boats around the Mediterranean. Back to bed, and up again at seven to fly the weather balloons.

He used the data to draw charts, and as the days went by stacks of useless weather maps, streaked with isobars and arrows, piled up around the barracks. He changed the name of the daily bulletins from "forecasts" to "now-casts," highlighting the difficulty of his task. Still, the Daedalus team demanded a more exacting science. Many nights he told them to be "on call" in case the winds shifted to a southerly flow or a high pressure ridge brought calm overnight. Many mornings, he awoke to placid, languid skies that he would judge too temperamental for flying.

Bussolari could still criticisms from the team, but he could do nothing to control rumors that moved out of the Xenia into Iraklion. One afternoon, Glenn came in from a training ride with news of the latest gossip. He'd overheard people in Crete saying the best weather had already passed. People believed the reason Daedalus hadn't flown was because the team wanted an American pilot. The Greeks suspected that Kanellos had been overlooked on purpose.

"You've got to admit we've had better weather than you can picnic in," Kerrebrock said. "We've got to give people a good, logical statement of what we're waiting for."

"Well, it could be that we have passed up the best weather," John said. "You know Bussolari did call a scrub one morning even with winds at three knots. We all agreed to wait for a better day. We understand why we didn't fly, but the locals don't."

"Sure, but even team members complain when they wake up in the morning and open their curtains and the sea looks like glass," Glenn said.

"You do have a working press center here," Brian Duff said. "Maybe you should open the weather briefings with a little more background."

"That might help with the weather," Glenn said, "but I'm hearing that people think there's anti-Kanellos sentiment on the team."

As usual, the meeting ended inconclusively. Two days later, on April 14, Kerrebrock and Duff would leave Iraklion and return to the United States. They couldn't wait any longer. Ethan would have to go soon, and so would Lois. One by one, professional obligations were starting to call home important members of the team.

Not long after the last staff meeting with Duff and Kerrebrock, an editorial cartoon appeared in a local daily newspaper lampooning the project. Drawn from an ancient Daedalus icon, the cartoon showed Daedalus sitting at work with a hammer, building a set of wings while his son stood and watched. The underline read, "It's a great shock, my son, in a world full of 'Gary Coopers' to discover that you are the only Indian. . . ."

Kanellos did not object to the cartoon nor did he issue a statement to the press. He knew his days, like the others', depended on the caprice of winds and rains. But he smiled faintly when he saw the drawing taped to a bulletin board at the Xenia. He said simply: "I am Greek. They are Greek. It is good to have support."

"John Langford! John Langford!"

Poutos stormed through the lobby waving copies of the morning papers. Editorials in two of the largest circulating daily publications in Greece had criticized the Daedalus Project, and one claimed that, according to a respected international newsletter, the project actually had disguised its real purpose. In its Tuesday, April 19, commentary, the newspaper *Ta Nea* reported charges from an article in the *Foreign Report* newsletter that Daedalus represented the first stage in the development of a new American defense-based spy satellite—"a spy in the sky," to be exact.

Poutos grabbed John and the director of the MIT news office, who had just arrived in Crete to replace Duff and Kerrebrock, and translated the editorials. John laughed; Poutos puffed furiously on a cigarette. Within hours, they found a copy of the original article, which had just appeared in *Foreign Report*, a weekly subscription-only publication of the respected British magazine *The Economist*. It described the team's weather delays and predicted that Daedalus technology could "lead to the production of unmanned high-altitude satellites that would be versatile, cost little and be virtually undetectable." And if there was any question about a hidden conspiracy within the dreamy, charming student-operated Daedalus Project, a backgrounding of the core engineers—whose jobs were linked to the American defense establishment—certainly seemed to provide evidence enough.

Poutos was most upset. While John typed a long, measured response to secretarial ministers to allay fears and deny allegations, Poutos spent the day in the press room answering phone calls from the government. Could it be true? Was Daedalus secretly tied to the American defense establishment? Poutos merely raised his eyebrows and shrugged. He grunted and groaned. He had his suspicions.

Kanellos seemed to enjoy watching John snared by the Greek press. Hours after the report appeared in *Ta Nea*, Kanellos predicted that he'd make the Daedalus flight the following weekend, during his rotation. "The plane will fly by Sunday," he said. "If not, I will call a press conference."

The next morning Juan discovered hairline cracks in the Daedalus A-ship's carbon-fiber tail boom. He found the fracture about a half hour before Frank was scheduled to take the airplane out for a trial flight on the runway. If the plane had flown, the tube probably would have snapped. Juan could neither identify the cause of the break nor tell when it occurred. He spent the day wrapping the boom, and expected to lose another day just rewiring control cables.

The pilots grew increasingly agitated over the delays. One of them complained to John that Glenn was gaining weight and that he'd started spending more time with Mary Chiochios than he did training for the flight. And no one had missed the chart tacked mysteriously to the bulletin board in the Xenia lobby, showing the results of Erik's latest heart-rate calibration test. It marked Erik's last bid to join the rotation, and by all accounts, it had appeared just before Ethan left for the United States.

In fact, Ethan *had* posted the measurements. The chart showed that Erik's physical condition had improved markedly over the past two months. He appeared to be just as healthy now as when he first joined the team in October. The message was clear: Let Erik join the pilot rotation.

But John held his ground. During debates before Ethan's departure, John again insisted that to enter the rotation, Erik would have to take a six-hour ergometer test, just as the other four pilots had done. Ethan argued that a six-hour test would sap the athlete's energy and put him out of commission for at least another two weeks. Time was running out. If Erik didn't get in now, in mid-April, he'd never fly.

Wednesday April 20 brought sweater weather. Juan greeted people at breakfast—"Surf's up!"—and motioned toward the whitecaps and spray of salt water splashing over the rocky breakwater. At the weather briefing, Bussolari, whose wife and infant had flown in from Boston, brought his baby to the press room while he drew the weather map. Juan sat down on the floor next to them and spoke to the child.

"Well, Nora," he said, "what's your excuse?"

The baby squealed.

"Two more days, huh?" Juan said.

Her father confirmed it. Same old news: X2. X2. X2.

Late that night, Erik and Frank walked into the Xenia lobby fuming. Juan and Mark and a group of UROPs looked up, startled at the sight of Frank, whose eyes looked sad and baggy. He had just spent an hour trying to persuade John to let Erik into the pilot rotation. Even Bussolari had given his support, but John still wouldn't bend.

"I don't think Erik being in shape is the issue," Mark said.

Frank sat down.

"No," Mark said, "the issue is what will the press say. John's worried about what the press would have to say about putting this guy in the airplane if he hasn't passed the six-hour test on the ground. The image of this project is basically everything now."

"Well," Frank said, "I just left John a note in his mailbox asking for a meeting in the morning. I'm giving up my place on the rotation."

Upstairs, in a temporary office, John was still debating how to handle the situation. He couldn't chance it. If Erik entered the rotation and then crashed the airplane because of a leg injury, a board of inquiry would be called—United Technologies had asked for reviews already. He imagined facing his peers and being asked why he'd used a pilot with a known injury who had not passed a standard test required of all the others.

That night John made perhaps his most unpopular decision as the Daedalus manager. Erik was out.

The deserted pilots' barracks where Bussolari developed his weather reports were quiet on Friday, April 22, except for the buzzing of a fly against a window screen. The day was warm and sunny. A wind sock just outside the door, which had flopped around like an elephant's trunk the day before, rested limply against its pole. The weatherman opened a jar of creamy processed apricots and started to feed his baby daughter. Two dozen maps of the Aegean, scribbled with wind direction symbols and the position of barometric pressure systems from the west coast of Spain to the east coast of Turkey, lay on a plain wooden table.

"I don't know . . . I don't know. . . ." He dipped a spoon into the jar and scanned the maps. "We've still got time."

The low pressure system over Turkey had, at last, dissipated. A large gradient of high pressure would soon ease into the Aegean. Nothing could disturb it until Monday. Saturday looked good. Sunday would be perfect. Would it be better to wait, or go tomorrow? He left Nora for a moment to check his computer for electronic mail from a private weather service in Boston. He called the meteorologist at the Air Force Base. He walked back to the table, gave Nora a cookie, and studied his maps again.

That night, before supper, Daedalus crew members slipped out on their balconies and looked toward Dia. A couple of UROPs had spent the day on mopeds exploring the Cretan coast and stopped at a church to pray for good weather. Another UROP, who spent his day on the dock near the Navy boats reading a book, reported that the winds there barely stirred his pages. Dari had taken a boat out to Dia just before supper for a measurement. No wind.

Bussolari promised a flight attempt Saturday morning.

"Initially," he said, "I expect to have a northerly flow of air in the morning at dawn. But once the axis of this high pressure system moves to the east we will see that unstable gradient turn to the south. At this moment, I'd expect that to happen Sunday morning. But we should take a hard look at tomorrow morning."

Nearly a dozen members of the team were already in Santorini. The miseries of politicking and waiting at the Xenia had left John with only a skeletal staff on Crete. But for those who remained to hear Bussolari's optimistic "now-cast"—not X plus two any longer—the words alone gave them a sense of freedom.

At ten o'clock, on the third-floor landing, an impromptu meeting was called for CPR training. The *National Geographic* photographer's wife, Tia, who had once worked as a nurse, assembled a group of engineers to instruct them how to save a drowning man's life. Peter Neirinckx, one of the UROPs, had Gup on the floor while the documentary film producer leaned down on Gup's chest and pumped. Peter held Gup's head back and pinched his nose.

"You have beautiful lips, Gup," Peter said.

"Just make sure you brush your teeth before tomorrow," Gup said.

"What do you do if he's bleeding?" Peter asked Tia. She tried to explain, but Peter couldn't quite take in all the details. "Just put pressure on the cut," she said, abruptly.

"Well, what do you do if he has a broken bone?" Peter pressed.

"That's the most likely problem," Tia said. "There's not much you can do. Just do your best." Tia was horrified by the lack of medical preparation.

"Does anybody know if Kanellos can swim?" Peter asked.

Greg Zack spoke up. "He's at least a good dog paddle," he said.

"But do you think he's likely to panic in the water?"

"Who?" Greg said, astonished. "Kanellos?"

At that moment a lean blond-haired figure appeared from the elevator smiling broadly. He lifted a couple of large plastic bags for everyone to see. Onions and garlic. Large, smelly, overripe onions and garlic.

"Steve Bussolari gave me this!" Kanellos said. "I will eat as soon as I land. I will keep the reporters away."

30

John awoke, startled at a light that flickered through the curtain of his hotel room. Just the porch lamp. It was 2:00 a.m., Saturday, April 23. He scrambled out of bed and went to the balcony. The sea looked calm. Bushes rustled below. Was it a land breeze, he wondered, or had a low pressure system formed during the night?

He showered and slipped a new blade into his razor. At two-thirty he left the dark lobby for the Air Force Base.

"Do you fly today?" asked the drowsy desk clerk.

"We don't know yet," John said.

He met Bussolari in the barracks. Detailed weather forecasts had just come from Boston; boats in the Aegean radioed reports of light winds. From his desk, Bussolari took calls from Jamie, Mary Chiochios, and Tim Townsend, one of the UROPs on weather rotation. They'd gathered readings all night on Santorini: Light winds everywhere.

John called the Xenia and alerted the team.

At four o'clock, as the lobby filled with crew members in orange Amoco raingear and as people wandered into the dining room to grab bread, eggs, cheese, pound cake, and bologna, Bussolari took readings for a final "nowcast." The Greek meteorologist at the Air Force Base arrived at the weather center, too.

"I think we have a day here," Bussolari said.

The Greek weatherman scanned his charts and went to the team's computer to look at other reports that had filtered in during the night. Winds blew from the north across the face of France; westerly winds drifted across Spain. While the weathermen conferred, a jet screamed across the runway outside. John glanced out the window and spotted the headlights of a Jeep illuminating the path of a narrow road that wound around the base to the barracks. Bussolari looked up. "That should be Lois," he said, "coming to mix the Ethanol." They stepped outside to check the red weather balloon. Monitors gauged the balloon's height at 50,000 feet. Winds aloft: zero.

The lobby of the Xenia emptied, except for a few stragglers. Finberg arrived late, all a'rattle, a tangle of wires poking out of his jacket like vines and two camera bags slung over his shoulders. He'd never managed to bring his autopilot to life, despite years of tinkering, but he'd arranged a deal with *Sports Illustrated* to shoot pictures of the Daedalus flight from a helicopter. A likeness of his head had been traced on a balloon and was hanging in the temporary hangar for good luck. Somehow Steve always had managed to be a harbinger of success, even if his electronics failed.

Kanellos sat at a table spread with a linen cloth. He wore red pants and a leather Daedalus jacket. He opened a box of Greek pastries—filo filled with nuts, oozing with honey—cracked open a loaf of whole wheat bread, and unscrewed the top off a jar of honey. He feasted alone, wide-eyed, wiping the sticky goo off his fingers.

"Just don't make us get up again," said the documentary film producer as he reached for a hard-boiled egg.

"No way tomorrow!" Kanellos said.

Peggie dropped by to leave the pilot a schedule of events. After the flight, he'd be expected in Athens for a series of interviews. He laughed when she disappeared out the door. He'd planned his own surprise for the media. He pulled down his warm-up britches and revealed black shorts riddled with holes. Saves weight, he explained. Kanellos would fly nearly naked.

When John stepped out of Bussolari's office he saw the temporary hangar lit up at the edge of the runway. Greek and American flags glowed through the hangar skin, and thin lines of the Daedalus airplane made their shadowy imprint on the hangar walls. There was Mark Drela in his orange Amoco suit. John counted the silhouettes of six others. He heard Gup call over a hand-held radio, reporting winds at the runway of four to five knots. Possibly a land breeze. It had been a northerly flow all night, not an acceptable pattern.

The water at the harbor reflected the red, white, and yellow lights of Iraklion, dissolving slowly like fireworks in a black sky. The Little Dipper hung over Dia. Empty fishing boats rocked at their moorings like cradles. A couple of UROPs spilled bags of ice into metal coolers to chill bottles of champagne. Soldiers on the Navy torpedo boat leaned silently on the railing of the deck, waiting for orders. Their colleagues' regular war games with the Air Force had been delayed, and their maneuvers had been moved more than ten kilometers to the west for the day to keep them out of the path of Daedalus.

John pulled into the parking lot at the harbor at 5:30 a.m. "It's dead calm all over the Aegean," he said. "Except at Santorini. It's started blowing like crazy there. We need a report from Dia."

He walked over to a Coast Guard boat and told the captain to prepare to leave the harbor in forty-five minutes. But the man didn't speak English. John waved and walked away. A radio call came from Bussolari—he wanted to check once more with the meteorologist in Boston to analyze the weather pattern. John jumped in his car and drove off again, out to the Air Force Base.

Erik was standing in the grass off the runway when John arrived, taking photographs for Reuters. A small crew of engineers, including Juan, Gup, and Mark, worked outside, piecing together the aircraft.

"I'm sick of this," Juan said, as John approached.

John didn't say a word. He simply reached out and shook Juan's hand. The propeller spun slowly as light gusts washed down the runway toward the sea.

Radio conversations blurred with the chatter of team members. Tim Townsend called from Santorini, referring to himself as "Double Tango," calling Jamie "Mountain Goat" and Mary Chiochios "Master Card." He radioed the command boat—"Double Tango to Papa Bear!"—and the Navy boat—"Double Tango calling Mama Bear!" Teammates in Iraklion who heard his transmissions looked at each other, mystified.

"Shut up!" Bussolari yelled. "Everyone shut up!"

At six o'clock the winds on Santorini measured five knots—the cut-off point. John returned to the harbor. Sunlight colored the seawater pink. Twenty minutes later, as he and his family boarded the command boat, the sea looked like a hot griddle. A rooster crowed and the song of terns overhead joined the thrum of engines. A fleet of Daedalus boats made its way out to sea.

"*Déjà vu,* heh, Yannis?" Bussolari said to the command boat's captain.

"This is it," he snarled. "No more days like this. This is my last time."

They heard radio reports from Juan and Gup at the runway. Three to four knots.

Kanellos had warmed up and was resting in the cockpit, waiting for the signal to fly. Bussolari poked his head out of the roof of the *Dulcy II* and, as the Daedalus boats bobbed in the waters at the edge of the cliffs, he ordered one of the runway crew members to light a smoke flare. John poked his head through the hatch, too, and rested his chin on Bussolari's shoulder as they both watched through binoculars. Smoke flowed off the cliff like a gray waterfall.

A report from Dia: winds were calm. All they needed now was a signal from Santorini. The five-knot northerly winds hadn't been acceptable earlier. From the command boat, they could barely hear Jamie's report. Heavy static garbled her transmission. Bussolari instructed the captain of the boat to turn them around 180 degrees to catch a better radio signal.

"Jamie," John said, "please repeat your offshore wind reading. Just the number, three times, slowly."

"One point four . . . one point four . . . one point four."

John looked at Bussolari. The readings were in meters per second, but the calculation was second nature—1.4 m/s = 2.8 knots, just within cut-off.

"Now the direction," he said. "Just the direction."

"Southwest . . . southwest . . . southwest."

It was a tailwind!

"We would like to proceed," Juan radioed. "Now!"

A few minutes after seven, the sun had turned a rosy red, and Bussolari gave the signal. "Okay," he said, "that's it. Let's take off."

"Let's get out of here!" John shouted into his radio. "Get out of here!"

From the command boat, voices of excited team members echoed vaguely from their distant vantage points up and down the shoreline. They could barely see the airplane as it lifted off the runway and seemed to rise even higher, not floating but hurtling through the air, growing larger as Kanellos pedaled over the cliffs. The long, long wings curled up like the last glimmering slice of an almost new moon, and then Daedalus came into view—

gigantic wings, but so thin that they almost vanished in the sparkling sun—and passed over their heads.

"Holy shit!" Tom Clancy yelled. *"Holy shit!"*

"It is better than perfect!" Kanellos radioed as he flew over the command boat. Boat engines growled and the flotilla fell in line. Daedalus flew one hundred thirty feet off the water toward Dia and set a course northward into the Aegean.

"Heart rate one hundred fifty," Kanellos reported. "Airspeed seventeen knots." He'd caught the tail wind.

For a half hour, the boats battled with each other for position to watch Kanellos. They ignored John's organizational charts establishing a proper alignment in the water. One inflatable was supposed to take the lead to give Kanellos directional reference and two others were supposed to be at each tip. Instead, they crisscrossed paths and skewed their assignments to watch the pilot pedaling at a fast clip, 105 rpms. As they passed Dia, Kanellos reported the altimeter had failed, but he flew so high off the water it hardly mattered. Inflatable boats, skipping like stones across the sea, zipped beneath him, and crew members craned their necks to watch. He rose to thirty feet, forty feet. In an hour, they lost sight of land.

At eight-thirty Kanellos finished his first liter of Ethanol. His heart rate dropped to 137 beats a minute. As he gave terse physiological readings over the radio, John and Barbara made notations on a clipboard. They had to keep him on a strict drinking schedule, as Ethan had advised. A liter an hour would stabilize his glucose level and stave off dehydration. But at this pace, with a three-knot tail wind and a stubbornly quick cadence set against the pedals, Kanellos had a chance to fly the path in four hours, not six. For long stretches passengers of the Daedalus fleet—Navy and Coast Guard boats, inflatables, Air Force helicopter, cabin cruiser, and a Cessna airplane—watched silently as the pink wings sliced the empty horizon across an endless sky. Like Daedalus, he flew neither high nor low, but unremarkably, the middle way.

John ate breadsticks, oranges, bananas, a half box of chocolate fudge cookies. He peered at Kanellos through binoculars, and from time to time barked at his crew members. But, at last, it was beyond his control. He left the cabin and walked to the back of the command boat. The captain killed his engines to dodge the wakes set by excited inflatable teams, then revved the motors again to catch up with Kanellos. He yelled at the students and leaned on his horn to get them out of the command boat's path. John laughed at the confused assembly falling behind the airplane. "This is nuts!" he said. "Totally nuts!" Ellis lay next to Barbara in the cabin, sleeping with a stuffed animal in his arms.

At nine o'clock, a sleek white fiberglass power boat cruised by to the left of the command boat. Juan stretched out on the deck with the wind whipping through his hair and uncorked a bottle of champagne. Mark leaned against a railing and faked a yawn.

"Congratulations, Kanellos," Bussolari radioed, "you've just broken the straight-line record for human-powered aircraft." But Kanellos didn't respond.

At nine-thirty, Kanellos reported he'd finished another liter of Ethanol. "Too slow," said Barbara, who'd been recording the drinking schedule on John's clipboard.

"My stomach is full and is sloshing around," Kanellos called.

A crew member in each inflatable aimed a video camera at the cockpit, and professional photographers collected images above and below Daedalus. They were a disposable event, a stunt. They'd created a fantasy so ephemeral and precarious that there would be no way to capture this again. Underneath, eyes trained down the long-limbed wings, so transparent that every white rib showed, so clean in their lines that the airplane seemed to swell or shrink, depending on whether one watched from the tips or the tail or under the small white pouch where Kanellos pedaled. They knew Daedalus too well not to sense the end of their journey the moment Kanellos had crossed the cliffs. They knew that Daedalus would make only one flight, and no more, that its wings flew at every moment on the verge of a stall, that for the ancient Greeks there had been little distinction between the sea and sky, and that in the vast blueness they had tethered together a machine that flew as a dream.

On the command boat, John grew restless. He paced. He made notations on his clipboard.

A television cameraman and producer had roped themselves to the bow of the command boat to maintain a steady position over four hours. A photographer for *Life* magazine, who'd waited four weeks in the Xenia's lobby for this day, mounted a lens that looked like a mortar shell on the front of his camera. After three years with the team, Chuck O'Rear, a *National Geographic* photographer, leaned out of the Cessna and fired through his last half-dozen rolls of film. How else would they remember what had often seemed so important? Hours passed immersed in the image. Even the sea looked smoky, like celluloid.

An enormous commercial cargo ship appeared suddenly in their line of sight. A call from the Navy boat warned that readings from the Navy's navigational compass showed that Daedalus had drifted 30 degrees off course. Bussolari repeatedly radioed Kanellos to correct to the left, and the captain of the command boat quickly punched buttons on his LORAN unit to cross-check the Navy's figures.

"The wake of that cargo ship is gonna cause real problems," Mark radioed. John and Bussolari knew immediately what he meant. For an airplane as sensitive as Daedalus, even the moderate air disturbances caused by a cargo ship's wake could tear apart the fragile foam wings.

A peppery burst of orders came from the Navy commander, and then the boat turned its guns in the direction of the freighter. Bussolari lifted his binoculars and howled as he watched the freighter make a sudden shift in course.

"I don't know what they told that guy," Bussolari said, "but he sure got a squelch on his headset."

"That's what Navy boats are for," John said.

The captain of the command boat didn't translate for them, but he clearly enjoyed the heated exchange. "The captain of that cargo boat," he said. "You should have heard him when he saw the airplane. He couldn't believe his eyes!"

Kanellos, still pedaling quickly above, sent a stern message to the team. "I do not like jokes now," he said.

The cargo boat turned 180 degrees and the flotilla slipped by undisturbed. They were certainly alone now. The boats fell into position, and because they were spread so far apart and Kanellos still had so far to fly, celebrations tended to be private and restrained. In those moments, they sipped champagne. Bryan Sullivan cried. Peggie stood motionless on the deck of the command boat. Gup fell silent at the helm of an inflatable. They stopped counting wing ribs and the revolutions per minute of Kanellos's cranking legs. They quit making calculations and monitoring the nearly imperceptible quiver of the spar. For once, they flew.

Shortly after ten o'clock, the dark shoulders of the Santorini cliffs appeared above the horizon. Nothing but a hint of land, more like smoke. The caldera seemed to rise up out of the sea slowly, an immense and startlingly dark visage. Everyone had imagined the flight out of Crete, but perhaps only Juan and the pilots had imagined the landing. John found himself gazing at the growing volcano that broke through layers of haze. Kanellos had pedaled for three hours, and for some reason, to John, it seemed far too soon to consider the end. He had flown more than 200,000 miles to make this flight. He'd lost friendships for it, he'd lost vital connections with his family. What would he find on the island now? What was there left to lose?

The command boat captain moved to the edge of his seat and leaned forward. At 10:20, Santorini was still just a vague outline, like a mirage over the water. Ellis woke up. Barbara kissed the little boy and tucked the stuffed animal—a gift from Parky—underneath his arm.

Bussolari lowered his binoculars. The black beach at Perissa came into view.

"John," he said, "Kanellos has so much left he could go to the airport and land at this point. I mean, we have a museum piece up there. We could avoid the beach crowds and save the plane. What do you think?"

"I don't think we should change course yet," John replied. "Besides, we don't even know where the airport is."

"I think he could find it," Bussolari said.

"Well, in that case, ask him if he wants to go to Ikaria."

Bussolari's head popped out above the cabin hatch and a moment later he looked back at John. "I asked," he said. "Kanellos says, 'No chance!' "

A few boats stopped a half mile from shore. Juan transferred into Dari's inflatable and Mark raced his boat toward the shoreline. The weather team at Santorini had driven down to the beach and radioed reports of slight one-foot swells just offshore and winds of 3.8 meters per second from south-southwest. Kanellos waved to the command boat from thirty feet, and eased off his pedaling so the airplane began to descend slightly. They could see red tan lines on his arms and heaving chest. The cliffs rising out of Perissa beach looked like elephant hide. Santorini came into relief and quickly overtook the horizon. The wind picked up. Radio transmissions bounced wildly between team members who'd raced ahead to prepare for landing, the lead inflatable boat where Bussolari plotted the airplane's final progress, and Kanellos, who could barely distinguish land and water through Daedalus' foggy windows.

John was only a spectator, with his wife and child, listening to the final spontaneous commands between boats, beach, and airplane. He could see Bussolari standing distantly at the helm of an inflatable, following right on the tail of Daedalus, shouting radio commands. At the base of the immense black cliffs, a blue and white sparkle caught his eye, a salty, gray sprinkling of homes and domed buildings along the beach. Daedalus was nearly invisible now. He sat on the deck of the command boat and listened to the radio.

There was no single commander anymore. Separated by necessity, his team guided Daedalus from a distance, blindly.

"We're watching your tail now," Bussolari told Kanellos. "Steve Darr, head for the radio tower on Santorini . . . Okay, Kanellos, your attitude looks real good. You're doing a good job."

"Bussolari, this is Steve Darr on the *National Geographic* boat. Okay, I want to give you an idea of what's going on at the beach. The Navy boat and the Coast Guard boat are moving to the southwest end of the beach. Kanellos will want to come in to their right. From your perspective, come to their right. The white NOVA boat will be anchored near there, so you will want to come in toward the white boat and make a sweep left away from it toward the Coast Guard boat."

"Thanks, Steve," Bussolari said, "I've got a power plant with a smokestack in sight and I'm assuming our landing sight is to the right of those. Right?"

"That's not a good assumption," Steve radioed. "The landing sight is probably three quarters of a mile to the northeast of that along the beach. So you'll want to come in much further to the right of the smokestack. Come almost directly or a little to the right of where the radio tower is on the mountain."

"Okay," Bussolari said. "Kanellos? Why don't you give me about three degrees of right wing direction? . . . Okay, nicely done. That looks good. Let's hold back for the moment."

John kept popping his head up into the hatch and down into the cabin to find the best angle. Bussolari watched the airplane's rudder shift to make corrections, and he took a deep breath before deciding how to direct Kanellos into the wind. The breeze had picked up so suddenly off the black sands that Daedalus wouldn't be able to go nose first. Any unexpected turbulence—a gust—would buckle the wings or break the boom and leave the airplane awash a quarter mile from shore.

"Okay, Kan," Bussolari said, "what we're going to do is descend once we get close to the beach. We'll make a descending left-hand turn. We'll turn to the left and come parallel to the beach. Once you're parallel, we'll want you to move to the right across the sand. We want to do that at an altitude of about a meter and a half and we'll be reducing that altitude the whole time. We want to cross the water and the edge of the sand very low to the ground. We want to be only about a meter off the ground, and then we'll want to do a very strong flair over a long period of time. We'll want to bring the nose well up—higher than we normally do it—so that we slow down a lot. Did you understand that?"

"Is okay," Kanellos said. "I understand. Now do we face the landing area?"

"No, we're not. We're heading to the right of the landing area, and I'd like you to head about three degrees more to the right."

"Okay," Kanellos said. "I cannot see very well, Bussolari. I have to be very close."

The Greek's legs flashed in the cockpit. The rudder whipped around and Daedalus made a sudden correction in attitude.

"Okay, fine," Bussolari said. "Now you have a defogging stick on your right-hand side that you can wipe the cockpit with. Can you do that now?"

"It's not working very well," Kanellos said. "There is a lot of fog on the screen."

"Okay, we'll just have to do it by feel. Come a few degrees more to the right."

The airplane climbed as Kanellos made the correction. His legs churned wildly.

"Bussolari, I do not have the altimeter," he called.

"I will try to give constant altitude calls as you're landing," Bussolari said. "I will try to give you as much information as you need."

"Bussolari? This is Mark Drela, over."

"Mark, did you call me? Over."

Mark stood on the beach listening to the exchanges on his radio. Kanellos was less than two hundred fifty meters offshore. They needed perspective from land.

"Yes," Mark said, "have him come in high. The wind is dead parallel. There is no danger of his overshooting the beach whatsoever."

"Okay, we'll come in high then," Bussolari said. "That sounds like a good plan. I'd rather him have the altitude . . . Okay, Kanellos, your heading looks real good at this time. And a landing party . . . Tim Townsend, I see the smokestack with the power plant just to the north of it, and to the northeast of it I see a big white building. Can you tell me between those two things where the landing site is? And, Mark, I would like you to release your first smoke bomb."

"Bussolari, I cannot understand well where the place you say is," Kanellos said.

"This is Double Tango. Could the white building you're talking about be the church at Perissa?"

"Okay," Bussolari said, "I have the Navy boat in sight. Could we have the first batch of smoke from the landing site? And, Kanellos, I want about five degrees to the right. Come about five degrees to the right."

"I can fire a flare now," Steve Darr radioed. "I don't know if it will help, but I can fire it up."

"Why don't you fire it up?" Bussolari confirmed.

"There's a flare up now," Mark said.

Even a few hundred meters out to sea the boatmen couldn't detect a line of smoke. They only saw crowds of people streaming down the beach.

"Kanellos, correct well to the right, come about twenty degrees to the right," Bussolari cried.

"He's dead parallel to the beach," Mark called.

"Come twenty degrees to the right, Kanellos," Bussolari said again. "Correct well to the right. You're heading way left of the landing site."

"The flare's going up," Steve Darr said.

"Center inflatable," Bussolari said, "I want all those boats out of the way! I want them well back! They can't be there. I'm trying to do a right-hand turn. Get them out of there!"

For the first time, John picked up his radio and called. "Bussolari, I'm worried about the beach. It looks like there's a shitload of people on that beach."

"Okay," Bussolari said. "Let's get the beach cleared . . . Kanellos, I want twenty degrees to the right. Give me a good right turn!"

"Okay," Kanellos said.

"Bussolari, all the people on the beach are with the project," Juan radioed. "There is very good crowd control here."

"Okay," Bussolari said, "I can see you guys now. Are you releasing the smoke at the inflatable? Okay, fine. I can see the smoke, lead inflatable. Are you in line with the landing site? Are you between us and the landing site?"

"That is correct," came a voice from the lead inflatable.

"Okay," Bussolari said, "I need another twenty degrees to the right."

"Bussolari, I am the only one on the beach with an orange boat suit." Barely discernible against the black cliffs, it was Mark, waving his arms.

The airplane struggled to turn into the wind. The rudder cocked hard against the tail boom.

"Kanellos, I need another ten degrees to the right!" Bussolari said. They were less than one hundred meters from shore.

"Okay, don't worry," Kanellos said.

"I never worry," Bussolari said.

"Is the landing area to the left?" Kanellos asked. The airplane had eased around so the cockpit wasn't visible any longer from the command boat. Kanellos was just a voice. The airplane moved as slowly as a cloud.

"Yes," Bussolari said. "The landing area is to your left. I don't want to start the turn yet. Wait to start your turn. We don't want to overshoot. You look good now."

"Double Tango," Tim Townsend said, "the Navy boat is going the wrong way."

Bussolari watched the Navy boat sidle in from the left, directly toward the landing site.

"Let's get that Navy boat the hell out of here!" Bussolari said. "We don't want him in the way. He's crossing our path . . . Kanellos, correct five degrees to the right . . . and one of you inflatables get that Navy boat out of the way!"

"Bussolari, this is Mark. Over."

"Okay, Mark."

"Bussolari, I am the only one on the beach wearing a bright orange boat suit. I'll be waving my arms like crazy. Why don't you look for that? It's more visible than smoke, I think."

"Okay, fine, Mark. I'll look for you," Bussolari said. "I would like smoke on the beach, though, just so if we get close I can see it . . . Okay, Kan, you're looking real good."

"Bussolari," Mark said, "the Navy boat is going to cut right in front of him."

"Yes," Bussolari said, "I'm telling the inflatable to sink him. Okay, I don't see smoke. I want to see smoke on that landing site. Light up a couple of them! I want to see more smoke than I've ever seen in my life!"

"Bussolari," Mark said, "with the wind here, the smoke's almost invisible. Trust me!"

At sea, the boat crews noticed an unexpectedly stiff breeze blowing, but nothing indicated winds that could extinguish a smoke flare. As the airplane neared the beach, though, they saw Kanellos struggling to propel Daedalus toward the shoreline. No one had planned for this.

"Okay, I'm looking for you, Mark," Bussolari said, "I don't see you. Do you see the airplane? Are we headed in the right direction?"

"You're headed right for me," Mark said. "The smoke flare's up right now."

"Okay, Mark, are we headed upstream from you or downstream from you? I got the smoke! Okay, I got the smoke. I got the smoke . . . Okay, Kanellos, we're gonna want to start the turn. It's gonna be a little bit tight. Not yet. Don't start the turn right yet."

"Bussolari," Mark said. "The turn will be . . . It's so windy here!"

"I got the smoke in sight," Bussolari said. "I can see where the landing site is . . . Okay, Kanellos, when I say to, I want you to turn. We're going to turn to the left. It's going to be a good turn. You're going to need to be—don't overbank it—but remember, it's going to be a good turn . . . I want all those inflatables out of the way!"

"Are we landing?" Kanellos asked. "Before this beach?"

"Land where the mini-bus is off to your left," Bussolari said. "You can probably see it. There are people running along the beach now."

"Bussolari," Mark called, "I'm waving the mast right now." Mark was an orange speck on the beach, and he held up a long straight rod from the Daedalus mast, whipping it in the air. Suddenly over the radio came a hearty laugh. It was Kanellos.

"I think there are a lot of people running!" he shouted.

"Okay," Bussolari said. "Don't worry. We're going to land where they aren't. So, okay, Kan, maintain a straight course right now. Okay, continue your turn. Continue your turn to the left . . ."

Inflatable boats kept crisscrossing paths to avoid the airplane as Kanellos edged closer to the beach.

". . . Okay, I can see the mast, Mark. I can see you waving the mast. Looks good . . . Okay, maintain that heading for now. Okay, Kan, you've got about fifteen feet of altitude. Let's descend now, reduce the altitude slightly."

"Bussolari," Mark said, "I would keep him there."

"Okay, Kanellos, keep the altitude and continue your turn. Okay, maintain a constant heading right there. You're looking real good. You're on base leg for a left-hand turn to final. Okay, Kan, I want a little bit of a left-hand turn, a little bit of a left-hand turn."

Incredibly, as the airplane approached the beach and slipped sideways with its pink right wing perpendicular to the sand, just twenty meters from the shore, gusts rose up under the airplane and pushed it back towards the sea. The pilot gripped his rudder control and twisted it to the full cocked position. He pedaled harder, and Bussolari gave quick directions to keep him from stressing the spars.

"Reduce power!" Bussolari said. "I want more left-hand turn. More left-hand turn!"

"There are big thermals here!" Kanellos said.

The airplane hung still in the air. Trapped in gusts off the beach, Kanellos kept pedaling as the propeller whipped around faster and faster and the rudder shifted one way, then another. But Daedalus could move neither forward nor sideways. Commands from Bussolari came rapidly, helter-skelter, yet nothing would nudge Daedalus any closer. The authoritative voice lost its edge and then dropped into a monotonous drone.

"More left turn . . . More left turn . . . More left turn . . . You're looking good. A little right rudder. Right rudder. Don't overdo it. Right rudder, right rudder, right rudder . . . Okay, you've got the wings level again. Level it out. Okay. Maintain that heading, maintain that heading . . . I want the wing runners right there. He's going to set down almost vertically. Okay. You're looking good, Kanellos . . . A little bit of a right-hand correction. A little bit of a right-hand correction. You need to keep the . . ."

Daedalus had zoomed in toward shore at over eighteen miles an hour when Aeolus slammed his fist down and said, no more. The pilot had been allowed near-perfect conditions for most of the flight, with a three-knot tail wind, skirting Air Force and Navy war games in the Aegean, pushing aside a cargo boat, soaring into Santorini, waving Daedalus' pink wings at the muscular volcanic cliffs. But gusts kept rising off the blistering black sand, and when those hot breaths slipped out into the Aegean, winds lifted up against the wings of Daedalus and matched Kanellos's power, knot for knot. As the plane stopped in the air and Kanellos pedaled harder, the right wing extended toward the beach but couldn't reach across the invisible, seemingly impenetrable boundary of air that divided land and sea. The wing shuddered, and then the airplane jerked. The nose dipped down and up and the fuselage wobbled. Peter Neirinckx, who was standing in a boat underneath aiming a video camera at the airplane, saw the motion through his viewfinder and then dropped the camera as soon as he heard the crack.

Bussolari said, "Uh-oh," the first time, when he saw the tail boom between the elevator and rudder snap and control lines break. A little light snap—and then Daedalus' long right wing lifted up, straight into the air, and it too snapped. Everyone stopped for an instant, and when Bussolari said, "Uh-oh," the second time, the huge wing flipped backward and Daedalus dropped like a pellet-stung bird.

Grant leaped into the water to rescue Kanellos and Dari immediately stood up in his boat, raising his arms to catch a falling left wing as the airplane plummeted. And then he was holding the right wing, too. One in each hand.

The pilot's face broke through the water. Kanellos sputtered. He was laughing.

On the command boat, fifty meters from the vision, John Langford stood gasping in disbelief. "I think he made it!" he shouted. "I think he made it!"

All the boats sounded their horns and camera crews came running. Bits of foam and Mylar floated up on the beach as Kanellos swam ashore. The Greek waved as crowds pushed in to greet him. Erik doused him with champagne, and Kanellos grabbed the bottle and splashed what remained over his friend. A piece of the wing lay on the beach, and despite the team's efforts to circle around the carcass, a dog pranced over the right wing and crushed it. Children grabbed hunks of pink foam for Kanellos to autograph. People scavenged for pieces of Mylar and broken bits of carbon tubes and stashed them in their pockets for souvenirs.

"We had a successful flight and a crash," John announced to reporters when he arrived at the beach. "Something for everybody. It's all there—Icarus and Daedalus. I'm in shock. It's amazing. I'm just in shock."

Yes, it had been quite a spectacle. One hell of an ending. A fantastic show.

EPILOGUE

They fished Daedalus from the water and dismantled the remains on the beach. Kanellos and Erik gripped each other around the shoulders and passed a bottle of champagne back and forth like sailors, stopping only to pose for the cameras. A line of soldiers kept the crowd in order. John, Bussolari, and Kanellos climbed atop the hull of a splintered orange fishing boat to take questions from journalists, who'd gathered in the sand. It wasn't John who pronounced the flight historic. John fulfilled his responsibilities by thanking all the sponsors, and then turned to Kanellos. Still wet from saltwater, sticky from a drenching of champagne, Kanellos hailed the flight as "a triumph for science, for man, and for history."

Within an hour, the spectators were gone. Island children traipsed off with autographs on broken hunks of pink foam. Kanellos was hustled into a helicopter for Santorini airport; he was due to fly to Athens. By noon, the remainder of the Daedalus team spread out on the sand, sharing a lunch of cheese and crackers, French chocolates and bottled water. There were no more intrusions. No more airplanes to stand between them and the embrace of a hot, unclouded sun.

Juan directed a crew of UROPs to haul the broken fuselage and spars to the trailer, hidden in a cluster of trees by the beach. He seemed strangely quiet. Then he wandered off. When he returned, alone, Juan disappeared into the dark trailer and dipped his fist into an end section of an unbroken spar. He dug out three handmade envelopes, each bearing his own carefully drawn Daedalus insignia and a sixty-drachma stamp. Six weeks earlier, he'd secretly concealed them in the tube at Hanscom, for once adding the weight of sentiment—one addressed to Bob Parks, the team's finest engineer; one addressed to Debbie Douglas, the Smithsonian historian who had become one of his best friends during the building of Emily; and one addressed to Al Shaw, the MIT lab instructor whose basement office and shops gave birth to featherweight airplanes, from Chrysalis to Daedalus. His air mail felt dry and smooth. He hid the envelopes in his jacket, left the trailer, and disappeared again into the trees.

The helicopter carrying Kanellos waited nearby for the project manager, but John vanished. He, too, had slipped privately into the back of the trailer to be alone with the plane. He stepped into the dark and examined water-sopped parts, shattered carbon spars, broken wings. Tears welled up, not for the loss of the airplane but for the end of the dream. Something perfect had perished on Perissa beach. With Daedalus destroyed, he wanted nothing more than to return to the beach and reclaim his place among the engineers. A host of obligations—to fly with Kanellos back to Athens, to conduct interviews, to attend requests from sponsors and MIT—waited for him outside.

Juan found John inside the trailer. Respectfully, he closed the door. Then John stepped outside, hugged Juan, and ran to board the helicopter. Sunglasses masked what he felt as he and Kanellos lifted up above the trees.

Less than an hour later at the Santorini airport, John and Kanellos sat on the grass waiting for the last helicopter flight to Athens. Neither man spoke. Buses from the beach arrived carrying reporters and a few crew members, Peggie, Barbara and Ellis, John Poutos. As they walked toward the terminal to wait for their own flight back to Iraklion, John looked at the waiting helicopter. He felt the same stirrings that had overtaken him inside the trailer.

The helicopter blades flinched, then revolved slowly. Someone called his name. John stood up to cross the tarmac, then looked back and saw Barbara and Ellis standing outside the terminal in a crowd of journalists. But this time John stopped. He told Kanellos to go ahead. John turned, and—for what felt like the first time in his life—he went the other way. He walked, then jogged. He tossed his briefcase in the grass, started to run.

"Barbara!" he shouted. "Barbara!"

And as he shouted for Barbara to wait for him at the edge of the runway, he plunged through the crowd toward Peggie. He grabbed her arm and tugged. "Come on!" he said. "Grab your stuff. Grab your stuff and come with me."

He hustled Peggie across the tarmac toward the helicopter. John told her the project was in her hands now. He helped Peggie into the helicopter, and looked across the seat at Kanellos. The Greek appeared astonished.

"You can handle it," John shouted. "Don't worry. You're wonderful!"

The director of the MIT news office leaned out of the door as John bowed down away from the helicopter blades and jogged back toward the grass.

"You can't do this!" he shouted. "You'll have to answer to Kerrebrock!"

"I can take it," John said.

Finally, the helicopter lifted up and disappeared over the black volcanic cliffs. John and Barbara and Ellis rested in the grass and watched the last airplane leave for Iraklion, full of writers and reporters and cameramen.

That night, forty-three members of the Daedalus team celebrated at a taverna in the town of Thira, high above the water. Waiters brought plates loaded with fresh fish and lamb, salads and pastries. The team drained a stock of Santorini wines. Glenn confessed that he had broken a rib water-skiing during their first week in Iraklion. He hadn't told anyone for fear that he would be bumped from the pilot rotation. Frank jumped up from table to table dancing to bouzouki music. They told stories of death pacts and kludges, Tools and hackers, slackers and Martyr Points. They relived the Daedalus crash in California and its final destruction in the Aegean.

At four o'clock in the morning the last members of the team left Thira and wandered off for their hotels. Nearly everyone gathered again at noon on a dock to take a boat back to Iraklion. It was a slow ride, and uneventful.

Juan's wish came true. Daedalus arrived home in a garbage truck. Scholars at the Smithsonian argued about its historical value and could not agree on its place in the pantheon of the world's aircraft. The pieces eventually went into one of the museum's warehouses to lie in the shadow of a decrepit World War II bomber until someone could marshal persuasive powers enough to call for an exhibition. So far, no one has.

The back-up airplane went on display for just one day at MIT's gymnasium, and then it was trundled into Jack Kerrebrock's backyard, where it sat for months in a trailer. Although the airplane had brought MIT more favorable notice than any single project in its history, Daedalus remained controversial among the engineering faculty. Some said the project had exploited undergraduate labor, and though Kerrebrock defended Daedalus to the end, even he seemed pleased when the hoopla finally passed.

Those who would look for a legacy could refer to a score of newspaper and magazine clippings—from the front page of *The New York Times* to a five-page spread in *Sports Illustrated*. NOVA aired an hour-long documentary called "The Light Stuff" during the fall of 1988. At the end of the year, a few Time-Life publications included a picture of Daedalus in their year-end roundups. But at the last, no story ever captured the spirit of the adventure. Daedalus hadn't been a triumph, as some suggested; the Daedalus project had never been about advertising or exhibitions. And no photographs ever portrayed the strangeness of the warm winds that rose so suddenly from the beach at Santorini and snapped the wings of Daedalus ten meters from the shore. Those weren't images for commercial consumption, as the engineers had sensed all along.

Of course they went on to other things. With the Daedalus Project finished, Parky turned his mind to a human-powered helicopter for a while. Gup returned to the West Coast, joined an expedition up Mount McKinley, celebrated Parky's thirty-seventh birthday in San Diego, and moved back to Boston. At last report, he was a graduate student in aeronautical engineering at MIT. A few UROPs built a human-powered hydrofoil with Mark Drela's help. Erik finished high in the Olympic trials in 1988, and Frank, who had privately struck a deal with Shaklee even before the Daedalus flight, managed the health company's cycling team for a season. He also ended the year with a steady girlfriend—Greg's sister. Greg, they say, has been searching for a pilot-training school. Glenn went back to medical school, expecting to pursue psychiatry, but found himself studying emergency medicine during the week and making weekend trips to Boston to visit Mary Chiochios.

After the flight Ethan lost interest in marketing the Daedalus endurance drink, and complained to Lois that age had finally hampered his own marathon training. He got married a few months later, and accepted a new job as director of the Pierce Foundation at Yale. Steve Bussolari returned to MIT to teach science. Peggie Scott moved to Atlanta. Steve Finberg went back to his old job at Draper.

Poor Juan searched without luck for an American airplane company whose lifeblood wasn't dependent on the defense industry. He looked for academic jobs in the United States, without success, and traveled through Italy and Germany, hoping to land a position with a sailplane manufacturer. At last, he took a job at NASA in Virginia working, happily, as a structures expert and taking courses in engineering.

The hopes John had held for so many years to start a company manufacturing solar-powered airplanes with Gup and Parky, Juan, Mark, and Finberg were never realized. But he did start a company, and he did begin plans to build a new airplane, one that would sample the ozone layer in the Antarctic for Harvard University-based scientists. He hired Tom Clancy and Siegfried Zerweckh, and hunted around rural Virginia for hangar space in which to pursue yet another dream. And only a few months later, Barbara gave birth to their second child, a son.

This is how it should end. Engineers return to institutions. Athletes join the field. The physiologist reenters his lab, and scholars continue their research. The airplane ceases to fly. When the project closed and all the equipment was sold, bartered or reclaimed, nothing remained but a catalogue of photographs, patches with meatball logos and icons of a team whose purpose now seems obscure. The images range across childhood, mock convention, and challenge the imagination. They celebrate the flights of youth and the will to dream. It should be a familiar tale, though it's not.

Daedalus, it's been said, worked for many different kings. He invented extraordinary tools and entertained children with his handmade toys. When he offended one king, he escaped to work for another. It was a hard way to go, living by his wits. He lost his child in the process, and eventually disappeared without a trace. But at least he saw a good bit of the world and, once, experienced the fullness of wings. Not a bad life, for a mortal.

AFTERWORD

In the end, it was Finberg who snapped the magic photograph that reduced to Kodachrome the whole three years of adventure and turmoil. From his noisy perch in the helicopter, he framed Kanellos sweeping into Perissa beach, with its steep volcanic cliffs and black sand, and the little town with its white buildings and blue-domed church. Other team members may grouse about the commercial nature of the way the picture was sold first to *Sports Illustrated* or that the image of the airplane is a little fuzzy, but for me that picture is the reason we did the whole project.

Well, maybe not the whole reason. There was a lot of childhood fantasy in it, too. I was only four years old when President Kennedy declared we would go to the moon, and I was six when he died. I was part of a generation raised on Walter Cronkite and *Life* magazine, and some of us grew up believing that if we studied hard enough and stayed determined enough, we could be in line when the torch was passed again. When Kennedy asked if we would accept challenges that would "organize and measure the best of our energies and skill," we prepared, and when he asked, "Will you join in this historic effort?", we signed up.

So it was a disappointment to come of age in a time when Jimmy Carter specifically offered "no new dreams" in his inaugural, and when the inspiration offered by Ronald Reagan was to "just say no." The development of the Daedalus Project probably drew at least part of its energy from this disappointment. Until our leaders offer enough vision to help us see beyond this week's popular crisis or next week's opinion poll, then I expect that people will continue to create their own small collective dreams. At least I would hope so. Adults, as well as children, need dreams.

We need dreams to sustain us. The extent to which we have lost the ability to dream is clear from the many people who have said to me since the flight, "It was too bad that it failed when it was so close to the beach . . ." How can you tell someone who didn't live through the experience, or understand it intuitively, that the flight wasn't a failure at all, that it literally was perfect? More perfect than we as mortals have any right to expect. To have the chance to fly, out of sight from land for hours in a pedal-powered fantasy, and then to see it all vaporize at the end, inches away from its destination, was such a poignant reminder of our mortality and humanity that it seemed scripted from atop Mount Olympus.

At the same time, the resolution of the project (as distinct from the flight) wasn't perfectly satisfying. The problem with dreams, as *Life* magazine once put it, is that they are doomed from the start. They either don't come true, which is painful, or else they do come true, which is fatal. Mission accomplished, end of dream. Despite its perfection, something died there on the beach at Santorini. Like any loss, getting over it takes time.

A few months after the flight, I gave a talk at NASA's Marshall Space Flight Center where Werhner von Braun and his team had brought forth the Saturn V moon rocket. As we stood in von Braun's old conference room and gazed out over the now-silent test stands an older engineer commented wistfully, "You know, we thought when we got to the moon successfully the nation would be grateful. Instead, they shut us down. Laid most of us off in months." I knew what he meant, because MIT did the same thing to Daedalus. Within a month of the flight, we were all off the MIT payroll, the solar home shuttered, the materials in the trash and the team dispersed.

I returned to my regular job in the real world with a mixed sense of bitterness and relief. The project had taken a heavy toll on my personal life and on my finances, and the plush, stable offices of a defense consulting firm near Washington provided a respite. But I soon found my mind drifting from the confines of newly prescribed duties. I kept thinking about the heights at Akra Spatha, its tenth-century ruins, the deep blue Aegean, and the timeless winds. Daedalus, they say, dedicated his wings to Apollo after his flight and never flew again. I have tried to deny it, but I never really liked the way that story ended.

I have left my job as a defense consultant. With a chemistry professor from Harvard as a partner, I have started a new company we call Aurora, named for the goddess of the dawn. Our offices are not nearly as plush as might be provided off the corridors of the industrial giants, but neither are they as sterile. As I write this we are at that wonderful stage where we have no money but feel the surge of idealism and see new aeronautical dreams forming, not around the sunny beaches of Crete but around the frozen deserts of Antarctica, where we hope to develop new techniques for researching the vanishing ozone layer.

I hope this afterword will be the last chapter for the Daedalus Project. But in making that wish, I have one final tie to make. This last task I do gladly: to personalize a concept that I struggled to explain throughout the project. It is the idea that, as one aerospace company's advertisement says, "No one flies alone." One of the real values of Gary's book is to make clear that despite our team's factions and fears and strong sense of individuality, it was a sense of shared vision and collective purpose that made Daedalus fly. Hackers, radical rocketmen, jocks, cynics and skeptics—this diverse group was united by the fact that each person was, ultimately, a dreamer and a believer. The tentative union of their individual dreams allowed a wonderful event to occur. Not all of them have been mentioned in this story, as the process of its telling required the author to highlight some people and not others. But I want to acknowledge them, and wish them luck too, as their journeys continue:

Jim Alman
Ed Becotte
Kevin Brown
Steve Bussolari
Mary Chiochios
Tom Clancy
Jean Cote
Juan Cruz
Jamie Dakoyannis
Steve Darr
Debbie Douglas
Mark Drela
Brian Duff
Steve Finberg
Jim Hilbing
Joice Himawan
Kanellos Kanellopoulos
Jack Kerrebrock
Eunice Knight
Theo Korakianitis
T.C. Lau
Lois McCallin
Sarah Morris
Jim Murray
Ethan Nadel
Peter Neirinckx
Bob Nesson
Mike O'Brian
Bob Parks
Konstantine Pavlou
Kelvin Phoon
Joanna Prentiss
Claudia Ranniger
Mark Schafer
Grant Schaffner
Eric Schmidt
Thom Schmitter
Frank Scioscia
Christine Scott
Peggie Scott
Tidhar Shalon
Al Shaw

Mike Smith	Bryan Sullivan	Matt Thompson
Louis Toth	Tim Townsend	Glenn Tremml
John Tylko	Craig Wanke	Dave Watson
Jim Wilkerson	Jonathan Wyss	Hal Youngren
Greg Zack	Siegfried Zerweckh	

Behind these stand others whose contributions allowed us to live out our dream. To me, they are all exceptional people, for each of them could have chosen to remain uninvolved. In most large organizations there is no penalty for a missed opportunity; the corporate structure, like society in general, makes it easy to just say no. So most people do. The few who do not, as Robert Frost might say, have made all the difference. They are represented for me in the figure of Ben Koff. Koff is vice president for engineering at the Pratt & Whitney division of United Technologies, and as such, he is something of a modern Prometheus, taking the fire of the gods and containing it in little metal cylinders that push those big aluminum buses around the sky. It was Koff who led Daedalus through the corporate labyrinth of UTC. He did so for education, to promote UTC's image among young engineers, and to give his company material, he once said, for classier ads. But the real reason, I think, is simply because Koff is a true engineer. In his veins flows something of the blood of Daedalus—part dreamer, part rascal, part artist. Our team understood this instinctively and responded to it. The very last thing added on the airplane before Kanellos's great flight to nowhere was a Pratt & Whitney eagle logo with its motto "Dependable Engines." This was not part of UTC's designer graphics, and they would have barred this addition had they known about it. But even as we were fending off pressure to put "one more MIT logo up front," this tribute to Ben Koff was applied without question or debate. This wasn't the logo of some soulless institution, after all. This was the icon of an engineer. He understood about dreams. I wish that there were more like him.

—John S. Langford
December 1989

FOR THE BEST IN PAPERBACKS, LOOK FOR THE [illegible]

[illegible]

[illegible]

[illegible]

[illegible]

[illegible]

In Australia: [illegible]

[illegible]

[illegible]

In Holland: [illegible]

[illegible]

[illegible]

[illegible]